PROJECT AIR FORCE

T0294879

Continuity and Contingency in USAF Posture Planning

Alan J. Vick, Stacie L. Pettyjohn, Meagan L. Smith, Sean M. Zeigler,
Daniel Tremblay, Phillip Johnson

Prepared for the United States Air Force

For more information on this publication, visit www.rand.org/t/RR1471

Library of Congress Cataloging-in-Publication Data is available for this publication.
ISBN: 978-0-8330-9565-7

Published by the RAND Corporation, Santa Monica, Calif.
© Copyright 2016 RAND Corporation
RAND® is a registered trademark.

Support RAND
Make a tax-deductible charitable contribution at
www.rand.org/giving/contribute

www.rand.org

Preface

In 2014 and 2015, the U.S. Air Force (USAF) published two long-range planning documents: a 30-year strategy (*America's Air Force: A Call to the Future*) and a 20-year Strategic Master Plan. The strategy

> provides a general path . . . to ensure our Air Force meets the needs of our great nation over the next 30 years . . . . This long look guides the 20-year *Strategic Master Plan*, which identifies priorities, goals, and objectives that align our planning activities with strategic vectors to produce a resource-informed 20-year planning force.[1]

The Strategic Master Plan has four annexes, including the Strategic Posture Annex.

To provide analytical support for the next draft of the Strategic Posture Annex, the USAF director, Strategy, Concepts, and Assessments commissioned a fiscal year 2015 RAND Project AIR FORCE study titled *USAF Global Posture and Presence*. The study developed a framework for long-term posture planning that accounts for both predictable and contingent drivers of USAF posture and identifies steps USAF can take to increase the robustness and agility of its posture over the 30-year planning period. The research described in this report was conducted within the Strategy and Doctrine Program of RAND Project AIR FORCE.

RAND Project AIR FORCE

RAND Project AIR FORCE (PAF), a division of the RAND Corporation, is the U.S. Air Force's federally funded research and development center for studies and analyses. PAF provides the Air Force with independent analyses of policy alternatives affecting the development, employment, combat readiness, and support of current and future air, space, and cyber forces. Research is conducted in four programs: Force Modernization and Employment; Manpower, Personnel, and Training; Resource Management; and Strategy and Doctrine. The research reported here was prepared under contract FA7014-06-C-0001.

Additional information about PAF is available on our website: www.rand.org/paf.

---

[1] USAF, *America's Air Force: A Call to the Future*, Washington, D.C.: Headquarters (HQ) USAF, 2014, p. 1. See also USAF, *Strategic Master Plan Executive Summary*, Washington, D.C.: HQ USAF, 2015.

Contents

Figures

Tables

Summary

Background

Posture planning is a juggling act. Because posture changes can be maddeningly slow to execute and, once implemented, difficult to change, planners must have a long time horizon. Good posture planning must distinguish powerful long-term trends from headline-grabbing but ephemeral events, have sufficient breadth to capture a wide range of possible posture demands, and be robust in the face of the inevitable uncertainties about where, when, and how U.S. interests will be challenged.

Meeting these requirements would be challenging even if planners faced a relatively bounded problem, such as "Design a posture in 2015 that will be able to meet U.S. requirements in Asia in the year 2030." U.S. Air Force (USAF) planners face an immeasurably more complex problem: Design a posture that can evolve to meet changing global demands over a multidecade period while making immediate adaptations to meet the urgent demands of today's crises and contingencies. Posture planners have to work simultaneously in three periods: the near, middle, and far terms. In this way, posture planning is similar to the challenge of sustaining and modernizing force structure: Current systems must be maintained to meet today's demands; programs must be developed, funded, and executed to modernize over the middle term; and research and development must push technology to meet long-term challenges.

Long-range planners and strategists recognize that future demands may be quite different from today's needs, sometimes in ways that would have been unthinkable a few years earlier. They account for these demands through trend and scenario analysis. These are valuable exercises. However, they are limited by contemporary intellectual frameworks and resource constraints, leading to a focus on the "most important" or "most plausible" challenges. Trend analysis is also constrained by human cognitive limitations (e.g., a preference for an orderly, understandable world; a tendency to overextrapolate current challenges into the future) that bias both consensus-based and top-down planning processes.

Although aspects of the future will be starkly different, it is simultaneously true that change by no means comes easily or uniformly. Long-range planning often neglects the forces that resist change and enable continuity in policy. Such forces can be powerful. On the one hand, they can strengthen USAF posture by making it more stable in the face of political or budgetary pressures or strategy changes that may be short sighted. On the other hand, these forces can inhibit efforts to fully align posture with strategy and may undermine posture agility.

Our analysis argues that planners must account for these diverse pressures as they develop posture in support of the 30-year USAF strategy. The dynamic posture model that we developed in this report offers a means to conceptually integrate and more deeply understand how these forces of continuity and change, deliberative planning, and randomness affect U.S. and USAF global posture. If USAF wishes to have a global presence that is robust, enduring, versatile, and agile, planners must account for and integrate these factors in posture concepts, plans, and programs.

Findings

Conflict Trends Are a Limited Guide for Long-Term Posture Planning

Recent trends in frequency, location, and type of conflict can offer insights for USAF planners regarding near-term demands. In contrast, conflict trends offer less utility for long-term posture planning. There are two reasons that this is the case. First, the most reliable long-term trends, such as declining death rates in conflicts and decreasing duration of conflicts, are not directly salient to posture planning. Frequency, location, and type of conflict are the variables most relevant for posture planners, but long-term trends are more difficult to discern. Perhaps most salient for USAF planners, decisions by U.S. leaders to use military force are not directly influenced by conflict trends. The future might, on average, have fewer and less lethal conflicts that suggest a safer world, but U.S. interests could still be threatened in significant ways, resulting in potentially large demands on USAF and other services. Alternatively, conflict trends overall might be on the rise, but the location and type of conflict might be perceived as either not directly harming U.S. interests or beyond the ability of the United States to resolve at a reasonable cost. Under these conditions, demands on USAF might decrease, particularly if U.S. leaders embraced a more restrained foreign policy or strongly resisted interventions in those types of conflicts or locations.

Unforeseen Crises or Conflicts Have Been Primary Agent of Change in Global Posture

Strategy is often assumed to drive plans, programs, budgets, and global posture. Historical experience suggests, however, that contingent events are the primary change agent in global posture. Strategy changes less often, frequently in reaction to unexpected crises. The major shocks that caused large increases, decreases, or shifts in overseas military presence were largely unforeseen: the beginning of the Cold War, the Vietnam War, the end of the Cold War, and the terrorist attacks of September 11, 2001. These events triggered fundamental changes in USAF global posture. In contrast, proactive changes in strategy (occurring before a precipitating crisis) are less common and typically cause only marginal changes in posture. For example, since the U.S. government announced a policy in 2010 to "rebalance" forces to the Asia-Pacific region, the Department of Defense (DoD) has shifted additional military capability to the Western Pacific and plans to do more in the coming years. Yet, even if turmoil in the Middle East and Europe had not slowed the strategy-driven posture shift, the change envisioned was likely to be small in both absolute terms and when compared to posture changes driven by major events,

ix

such as 9/11. This is not to suggest that the Pacific rebalance is mistaken or of no consequence, only that the scale of change is not comparable to that caused by the events listed above. Moreover, it has proven difficult to execute strategy-driven realignments in periods of strategic ambiguity and absent a compelling and urgent need. That said, if the United States moved from its grand strategy of deep engagement to one of offshore balancing (with minimal forward presence), the posture ramifications would be huge. A change in grand strategy of that magnitude, however, is rare and would likely only happen in response to some major stimulus— which again suggests that contingent (and often unexpected) events are first among equals as driving mechanisms for posture.

Posture Characterized by Long Periods of Stability, Punctuated by Rapid Change

Once the United States establishes large bases in a country or region, they tend to remain for extended periods. Multiple U.S. military facilities in Asia, Europe, and the Middle East date back to the 1940s and 1950s. This is primarily due to the phenomenon of path dependence, which creates positive feedback loops to maintain and enhance posture at the individual base, country, regional, and even global level. Path dependence had a powerful effect on USAF posture during the Cold War. Although marginal adjustments are made to posture on a routine basis, it has required a major international event to break down the stabilizing influence of path dependence and create the institutional momentum for large posture changes. These posture "shocks" happened at the beginning of the Cold War, beginning and end of the Vietnam War, end of the Cold War, and after 9/11. In the post–Cold War era, shocks have become more common, resulting in less stability. One can see the continuity and change in USAF posture by examining the proportion of bases by region.

The majority of USAF overseas bases were in Europe for most of the Cold War, except during the peak years of the Vietnam War, when just over 50 percent of USAF bases were in Asia. With the end of the Vietnam War and the drawdown in Asia, Europe once again possessed the majority of USAF bases. This was to be expected, as the major military threat to U.S. interests was in Central Europe. Yet most USAF bases remained in Europe well after the Cold War ended, suggesting path dependence played a role as well. This situation held until 9/11, when USAF presence in the Middle East greatly expanded and surpassed that in Europe. Yet by 2011, Europe once again hosted the plurality of USAF bases, suggesting that USAF presence in Europe is the stable equilibrium that the system returns to after disruptive events.

USAF posture probably will be more volatile in the future—similar to the greater volatility of the post–Cold War era. This is likely to be the product of a number of factors, including more-varied and geographically dispersed threats, efforts by DoD to rely on "lighter" facilities, and changes in collective beliefs about the legitimacy or appropriateness of U.S. bases overseas. To date, the first factor—the changing international environment and particularly the location of security challenges—has accounted for the majority of posture revisions. Despite DoD's emphasis on "places not bases," this approach has largely failed to prevent the United States from being locked into a particular base; many "lily pad" bases seem nearly as resistant to change as main operating bases (MOBs). It appears that path dependence in U.S. posture has more potential to be undermined by a collective backlash against U.S strategy or a more general pushback against the appropriateness of foreign bases. While the factors discussed above will probably lead USAF to make more changes to its posture, especially in response to future shocks, path dependence also appears likely to provide a certain amount of stability to its core base network (large bases with a continuous U.S. presence).

Path Aversion Can Constrain Posture, but Effects Tend to Be Fleeting

Path aversion, the process in which American leaders, policy elites, or the public resist particular types of military operations, is a less common but nevertheless important constraint on USAF posture. It is most visible when U.S. leaders decide to draw down deployed forces and associated posture prior to the full achievement of U.S. objectives, as they did in Vietnam and Thailand (1969–1975), Somalia (1993–1994), and Iraq (2007–2011). Path aversion following the Vietnam War triggered fundamental institutional reactions as well, including the Congressional War Powers Resolution of 1973 and U.S. Army doctrinal and force structure changes. Finally, path aversion following Vietnam (particularly in the Congress and public) limited President Ronald Reagan's use of force in support of El Salvadoran government counterinsurgency (COIN) operations in the 1980s, likely preventing the development of posture in El Salvador that would have been necessary if U.S. combat forces had directly intervened. That said, Vietnam-related path aversion seems to have softened within a few years as a constraint on presidential use of military force (aside from El Salvador, which for many looked too much like Vietnam for comfort). For example, President Jimmy Carter ordered the Iran hostage rescue in 1980, and Reagan ordered ground forces into Lebanon and Grenada in 1983.

More recently, path aversion toward ground-centric COIN operations in the greater Middle East dominated between 2008 and 2014. Today, policymakers and the public appear torn between a desire to see an end to U.S. combat operations in the greater Middle East and a belief that the Islamic State of Iraq and Syria (ISIS) represents a form of violent extremism that cannot be ignored and might even threaten the United States itself.[2] Path aversion, however, has eroded significantly since 2014; as of this writing (April 2016), the U.S. ground forces' role against ISIS has expanded to include artillery support and special forces raids. It remains to be seen whether path aversion in the Middle East will be a powerful or lasting constraint on USAF posture. Because ground-centric COIN is most strongly associated with Iraq, path aversion along these lines has arguably increased demands on USAF posture, since the campaign against ISIS (to date) has been air-centric.

---

[2] The terrorist attack by ISIS sympathizers in San Bernardino, California, raised concerns that the United States might be vulnerable to ISIS-inspired or ISIS-directed attacks. See Lara Jakes and Dan De Luce, "Is the San Bernardino Attack the Latest in 'Crowdsourcing' Terrorism?" *Foreign Policy*, December 3, 2015.

Geography Remains Relevant to Long-Term Posture Planning

Our assessment of 21 distinct planning vignettes (randomly selected from a population of about 150 vignettes) found that airfields in Europe and the Middle East were used in more operational vignettes than any other region. Airfields in North Africa are well located, but lack the infrastructure to support the missions required in the vignettes. Airfields in Asia were used less frequently, as relatively few Asia scenarios appeared in this one random draw and most Asian airfields are too remote or widely separated to be of use across regions. This finding is consistent with the findings of a 2013 RAND Project Air Force study that assessed the global utility of 30 airfields across several dozen scenarios in 12 distinct regions.[3] Bases in the Mediterranean littoral (Europe, North Africa, the Middle East) are uniquely versatile because airfields in any one of these locations can be used to meet many operational demands in all three. Bases in Europe are also critical en route stops for airlift to the Middle East and Africa.

Recommendations

This research leads to three recommendations for USAF leaders and posture planners.

Test 30-year posture plan robustness against the failure of key assumptions and across a large number of possible demands. Although the future will be a combination of the familiar and the alien, there is no definitive way to determine which factors will change and in what degree. Planning processes typically are based on assumptions, making judgments about the relative probability and importance of demands and other factors. Some, perhaps many, of these assumptions are bound to be wrong. To account for the irreducible uncertainties about the future, the planning process should seek to reduce the importance of assumptions by designing a posture that is robust across many alternative futures, including diverse assumptions and a wide range of demands.

Supplement operation plan–based posture analysis with broader consideration of future demands that current planners may consider implausible, unimportant, or both. Planners in 2016 cannot know with certainty what problems USAF leaders will face in 2044 (the last year covered by the 30-year strategy). Neither can they know which of these problems will be policy priorities. Two techniques can help reduce the impact of today's cognitive and institutional biases on long-term planning: massive scenario generation and contingent event analysis. The former uses simulations to consider the implications of hundreds or thousands of possible futures. Massive scenario generation is a relatively new technique that requires significant computational power but has great potential. Contingent event analysis is a simpler and less–resource intensive technique that randomly draws scenarios for analysis from several hundred scenarios that reflect a wide range of operational and geographic demands. We

---

[3] Stacie L. Pettyjohn and Alan J. Vick, *The Posture Triangle: A New Framework for U.S. Air Force Global Presence,* Santa Monica, Calif.: RAND Corporation, RR-402-AF, 2013, pp. 30–35.

recommend that USAF long-range posture planning incorporate one or both of these techniques to test future posture options across a wide range of demands.

Seek a balance between "stickiness" and agility in USAF posture. When compared to the demands of a given moment, USAF global posture has always had shortfalls and excesses. Yet when viewed over a longer time horizon, this posture has proven to be highly robust. It would be considerably less robust if it were easy to close MOBs and move permanently stationed forces based on short-term understandings of optimality and military demands. MOBs and permanently stationed forces make contributions beyond those associated with host-nation security. They also are critical to USAF global mobility en route structure. MOBs, such as Ramstein, Kadena, and Yokota, are all key parts of this network. MOBs also act as hubs from which permanently deployed forces can deploy to expeditionary locations elsewhere in theater or in nearby theaters. Thus, MOBs are often key enablers for USAF deployments to a wide range of less-developed facilities. The positive side of path dependence can be seen in USAF in Europe. USAF (and all military) force posture in Europe faced considerable pressure to reduce only a few years ago; now force posture is being expanded to meet an increasingly aggressive Russia. Path dependent processes, although by no means uniformly good, have helped enhance stability in USAF posture. That said, USAF also needs agility in posture, particularly with respect to demands in areas where enduring posture is neither politically feasible nor necessary. The ideal posture would combine sufficient stickiness to maintain enduring access in key locations and sufficient agility to surge from these locations to meet out-of-area demands and shrink back as operational demands end. The recent practice of developing small, scalable facilities with partner nations for use in exercises, training, and operational rotations is intended to increase agility, but as noted in the findings, can fall prey to stickiness as well. Creating and maintaining a posture that has elements of both stability and agility will therefore require active management by senior leaders.

Acknowledgments

The authors wish to thank study sponsor Maj Gen David Allvin (then director, Strategy, Concepts and Assessments, Headquarters [HQ] USAF) for the opportunity to consider posture demands over a long planning horizon. We thank Nancy Dolan (deputy director of Strategy, Concepts and Assessments, HQ USAF) for her helpful comments on this work. Col John Trumpfheller, Scott Wheeler, Ericka Reynolds, and Julie Birt in Strategy, Concepts and Assessments, HQ USAF provided constructive feedback and assistance over the course of this study.

At RAND, we thank project member Kathleen Reedy for her related research and colleagues Paula Thornhill and Lara Schmidt for their help with study design and methods. Lt Col Michael Jackson, RAND USAF fellow and project member, made important contributions, particularly with respect to airfield requirements and special operations demands. Reviewers Andrew Yeo of Catholic University and Bryan Frederick of RAND provided insightful and constructive recommendations that greatly improved the report. The authors also thank Col Michael Pietrucha, USAF, for his careful reading of the manuscript and detailed suggestions. Rosa Meza provided administrative support for project meetings and travel and prepared the draft report. Finally, thanks to our editor, Jessica Wolpert, for her skillful edit of the final manuscript.

Abbreviations

AAFIF	Automated Air Facility Information File
COCOM	combatant command
COIN	counterinsurgency
CSL	cooperative security location
DoD	Department of Defense
EAF	Egyptian Air Force
ESAF	El Salvadoran Armed Forces
FOS	forward operating site
GDP	gross domestic product
GDPR	Global Defense Posture Review
GTD	Global Terrorism Database
HQ	headquarters
ISIS	Islamic State of Iraq and Syria
ISR	intelligence, surveillance, and reconnaissance
JCS	Joint Chiefs of Staff
JTAC	joint terminal attack controller
MERS	Middle East Respiratory Syndrome
MKAB	Mihail Kogalniceanu Air Base
MOB	main operating base
NATO	North Atlantic Treaty Organization
NSC	National Security Council
PCN	Pavement Classification Number
PLO	Palestine Liberation Organization
PRIO	Peace Research Institute Oslo
RAF	Royal Air Force
RPA	remotely piloted aircraft

SAC	Strategic Air Command
SLIC	slow-intensity conflict
START	National Consortium for the Study of Terrorism and Responses to Terrorism
UCDP	Uppsala Conflict Data Program
USAF	U.S. Air Force
USAFE	U.S. Air Forces in Europe
USEUCOM	U.S. European Command
USMC	U.S. Marine Corps
USN	U.S. Navy

1. Introduction

Background

As U.S. Air Force (USAF) leaders work to craft a global posture (i.e., overseas forces, facilities, and arrangements with partner nations) that can meet the demands of today and those of the next 30 years, they must wrestle with a simple truth: In many ways, 2044 will be very similar to today, while in others, it will be jarringly alien. The problem for planners is that no one can predict where there will be continuity as opposed to change. To put this in more concrete terms for USAF posture planners, some, perhaps many, existing USAF overseas bases likely will be part of the 2045 global posture, and some current demands on USAF will remain as well. That said, if history is any guide, USAF likely will be called on to conduct missions and operate in ways and locations that appear today as not just improbable but unthinkable.[1] The challenge for USAF is to find a planning framework that integrates both stasis and change.

Although planners focus on predicting and adapting to new demands and constraints, they also need to account for forces that resist change. Forces that support the status quo are found in a host of areas: history, cultural norms, political processes, economic incentives, organizational cultures and procedures, and physical infrastructure, to name a few. A simple example of stability in global posture is USAF's continuous use of Kadena Air Base on Okinawa for over 70 years, a remarkable time span. Initially captured in 1945 during the U.S. invasion of Okinawa, the base was rebuilt and greatly expanded, initially to support the postwar occupation of Japan. Kadena was used during both the Korean and Vietnam Wars, was a key Cold War base, and remains the most versatile USAF base location in the Western Pacific.[2] An understanding of the forces and factors that lead to such an enduring presence at a location, in a country, or in a region is just as critical to effective long-range planning as creatively imagining alternative futures. This is a major theme of our analysis and will be discussed in detail in Chapter Five.

Long-range planners must also account for the likelihood that, in any three-decade period, there will be at least one major and unforeseen event that acts as a shock precipitating significant changes to the U.S. global posture. For example, military planners working on creating a post–World War II basing plan in 1945 did not anticipate the Cold War, let alone the Vietnam War, the Gulf War, or the terrorist attacks of September 11, 2001—all significant events that affected the U.S. military presence overseas. Unexpected shocks are inevitable over any long planning

---

[1] It is useful to distinguish between unexpected events in unexpected locations and unexpected events in locations where the United States has established bases. The former tend to be greater shocks to posture than the latter. The authors wish to thank Michael Pietrucha for this observation.

[2] Wesley Frank Craven and James Lea Cate, *The Army Air Forces in World War II:* Vol. Five, Washington, D.C.: Office of Air Force History, 1983, pp. 691–693; Kadena Air Base, "Kadena Air Base History," webpage, n.d.

horizon. Sometimes the shock is caused not by a new demand but by the end of an enduring force planning requirement. This is exemplified by the end of the Cold War. In 1985, well before the end of the Cold War, there were 27 major USAF bases and 87,765 airmen in Europe.[3] At the time, few foresaw the end of the Cold War or anticipated the U.S. military presence in Europe plummeting. Yet, 30 years later (in 2015), only nine major USAF bases and 28,920 airmen remained in Europe—approximately one-third of the 1981 numbers.[4]

Although the beginning and end of wars (both hot and cold) are arguably major sources of "posture shocks," lesser international events and crises, such as those during 2014–2015, can nevertheless disrupt posture planning. As recently as 2013, there was (1) considerable pressure from Congress to substantially reduce the U.S. military presence in Europe; (2) an expectation that U.S. posture in the Middle East and South Asia would shrink as the wars in Iraq and Afghanistan ended; and (3) a growing consensus within the Department of Defense (DoD) that U.S. posture in East Asia needed to be enhanced, adjusted, and expanded. Posture planners were busy adapting to these demands when, in 2014, Russia annexed Crimea and invaded Ukraine. At roughly the same time, Islamic State of Iraq and Syria (ISIS) emerged as a major policy concern for U.S. leaders. Russia's bold and aggressive actions and the success of ISIS disrupted the planned rebalance of U.S. posture away from the Middle East and Europe and toward Asia. As of 2016, U.S. European posture is being enhanced (primarily through infrastructural improvements, prepositioning of equipment, and rotations of forces), while the Middle East posture is creeping upward. Previously planned adjustments to Asia posture are continuing (e.g., basing of littoral combat ships in Singapore, a growing U.S. Marine Corps [USMC] presence in Darwin, Australia, and various base construction projections in Guam), but it would be inaccurate to describe these relatively modest initiatives as a significant pivoting or rebalancing as it was understood five years ago.

Posture planners face a dilemma in designing a posture that is robust in the face of such shifts in demand occurring over multiple decades. On the one hand, current security and fiscal concerns drive the posture conversation in Washington, pushing policy to optimize for the near to middle term. On the other, developing a robust long-term posture requires a dynamic consideration of how today's challenges may evolve and a recognition that the future holds surprises that will likely place entirely different demands on USAF posture.

---

[3] Major bases include only operating airfields that host USAF aircraft. Stacie L. Pettyjohn and Alan J. Vick, *The Posture Triangle: A New Framework for U.S. Air Force Global Presence*, Santa Monica, Calif.: RAND Corporation, RR-402-AF, 2013, pp. 67–68.

[4] Defense Manpower Data Center, "Active Duty Military Personnel Strengths by Regional Area and by Country," data set, March 31, 2015. The nine bases with flying units are Incirlik, Spangdahlem, Ramstein, Royal Air Force (RAF) Mildenhall, RAF Lakenheath, Aviano, Sigonella, Papa, and Geilenkirchen. The last two are North Atlantic Treaty Organization (NATO) bases, but they house NATO flying units that include USAF participation. Sigonella is a U.S. Navy (USN) base, but it is used by USAF intelligence, surveillance, and reconnaissance (ISR) aircraft. RAF Mildenhall is slated to close. USAF retains access to but does not continuously have flying units at RAF Fairford, Lajes, Morón, and Rota.

Policy Problem

USAF has identified a need for a 30-year posture plan that (1) addresses near-term demands, (2) can evolve in response to changing conditions in full recognition of the irreducible uncertainties regarding future demands, and (3) is sufficiently robust to meet USAF demands when inevitable surprises and "posture shocks" occur.

Purpose of This Document

This report is intended to provide an analytical framework and recommendations to inform ongoing revisions to the USAF Strategic Master Plan Posture Annex. In particular, it makes recommendations regarding the steps USAF can take to develop and sustain a global posture that is characterized by stable and reliable access and by agility.

A Simple Model of U.S. Global Posture

We argue in this report that four potential mechanisms drive future U.S. global posture: strategy, contingent events, path dependence, and path aversion. Each mechanism captures a unique dimension of the process by which posture is built, sustained, or dismantled. These driving forces may coexist as opposing pairs. Strategy reflects deliberate planning in which posture is intended to support long-term U.S. interests and strategy choices. Opposing strategy are contingent events that can trigger major changes in posture. Contingent events may be completely unexpected (e.g., the 9/11 attacks) or considered probable at some point but unexpected in specific manifestation and timing (e.g., war with Japan in the 1940s). Contingent events also lead to policy decisions, the ramifications of which are rarely fully understood when made. Path dependence helps explain the resistance of posture to change. Opposing it is path aversion, a process in which a bad experience leads to a desire to avoid entire classes or regions of conflict. This is exemplified by what has become known as the Vietnam Syndrome, a roughly decade-long period after the fall of Saigon in 1975 during which U.S. leaders and the public were loath to commit ground forces in peripheral conflicts.

Figure 1.1 presents these four driving mechanisms in a dynamic model. The idealized view of posture as strategy driven is captured in the upper left corner. Planners identify enduring U.S. interests and generate a national military strategy to protect these interests. Overseas bases are developed and forces deployed abroad to support the strategy. Although appealing conceptually, it is difficult to identify actual cases in which posture was created and developed absent compelling immediate needs. Rather, the spark for posture development more typically starts with a series of contingent (often unexpected) events in the form of internal political changes in host nations, crises, or conflicts, as shown in the lower left corner. These events may lead immediately to a realigned posture (e.g., when a new host-nation regime directs the United States to leave a country on relatively short notice), to changes in strategy, or directly to policy and posture actions (e.g., in 1990, when the United States rapidly expanded its Persian Gulf posture in response to the Iraqi invasion of Kuwait).

Figure 1.1. Dynamic Model of U.S. Global Posture

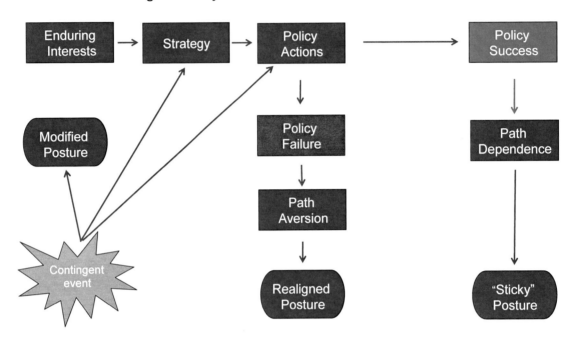

These policy/posture actions may succeed or fail. It is difficult to identify purely objective criteria for either success or failure. We define success as those situations in which actions create self-reinforcing processes, such as increasing returns, within a broader policy context that is viewed as working (e.g., development of NATO posture during the Cold War). Failure is defined as situations in which policy is no longer perceived as likely to lead to the achievement of U.S. objectives, at least not at an acceptable cost.

When policy actions succeed (as shown in the upper right green box), they can set in train a process that rapidly locks in a particular policy or path. This path dependence results in a "sticky" posture—one resistant to change.

Alternatively, if the policy fails, it leads to path aversion, a process in which U.S. leaders and the public are hesitant to engage in certain types of operations. This can be manifested in several ways. During a failing operation, leaders may decide to draw down forces and posture prior to any clear resolution of the conflict. This happened in Vietnam between 1969 and 1973 and in Iraq between 2007 and 2011. Alternatively, after the clear end of a failed operation, there may be additional policy and posture changes. Finally, path aversion can affect future decisions about intervention, leading U.S. leaders to limit or avoid prospective operations and not develop basing that would have otherwise been necessary to support the military mission. Path aversion is relatively rare in U.S. foreign policy, although the most famous case—captured in the slogan "No more Vietnams"—haunted U.S. policymakers for decades.

4

Postures resulting from the various path dependent processes may change on the margins, but they are remarkably stable for long periods—until a disruptive event triggers a major reset of posture in and across regions.

Organization

Chapter Two describes and analyses long-term conflict trends to provide context for our exploration of the four driving mechanisms for global posture. Although conflict trends are of limited value in understanding long-term posture requirements, elite and popular perceptions about these trends powerfully frame policy discussions. Because elite and popular perceptions have, at times, been at odds with the empirical evidence, we felt it important to address this debate before presenting our posture model in detail. Chapter Three introduces strategy as a potential driving mechanism for posture and assesses the posture implications of three alternative grand strategies. Chapter Four (1) argues that long-term posture planning must include the consideration of contingent events that current planners consider neither plausible nor important, (2) describes our method of accounting for contingent demands, and (3) presents results from our analysis of randomly chosen operational demands. Chapter Five introduces the idea of path dependence and uses posture data from 1953 to 2011 to assess whether path dependence occurred in USAF posture during that period. Chapter Six introduces the idea of path aversion, depicts historical examples, and describes the current situation, in which path aversion of Iraq-like conflicts has been overwhelmed by the problem of ISIS. Chapter Seven presents study conclusions and recommendations. The appendix describes the use of a Polya Urn model to gain additional insights into path dependent processes.

2. Long-Term Conflict Trends

As USAF considers potential demands on posture over the next three decades, it is instructive to consider whether there are long-term trends regarding the frequency or nature of conflict. Some argue that the world is more violent and more dangerous today than at any other time since the end of World War II. Conflict anywhere in the world certainly has the potential to destabilize countries and possibly larger regions. But are conflicts and violent acts more prevalent now than in prior decades? Is there something uniquely precarious about the security environment in the world today? These very questions have engendered a very public discussion—one largely pitting policy and defense practitioners against academics—about violence and the current state of the world.

Former Chairman of the Joint Chiefs of Staff General Martin Dempsey has argued on multiple occasions that the world is increasingly violent and unstable. For example, in 2012, he observed "that in my personal military judgment, formed over 38 years, we are living in the most dangerous time in my lifetime, right now." The following year he reiterated this view: "I will personally attest to the fact that [the world is] more dangerous than it has ever been."[5] While some may view Dempsey's comments as reflexive threat inflation in the face of budget austerity measures, others agree with him. Senator John McCain echoed Dempsey's sentiment during a 2014 interview with CNN, saying that the world was in greater turmoil than at any other time in his own lifetime.[6]

In contrast, political scientist John Mueller has argued, "[In] some very important respects, the institution of war is clearly in decline." Mueller's general evaluation is that war has been reduced to its "remnants," consisting primarily of its residual combatants, who are mostly participating in unconventional civil wars. More recently, psychologist Steven Pinker made the case that "we may be living in the most peaceable era in our species' existence."[7] Exploring the psychology of violence and nonviolence, Pinker propounds that violence of various scales—in

---

[5] Martin Dempsey, "FY13 National Defense Authorization Budget Request from the Department of Defense," testimony before the Senate Armed Services Committee, Washington, D.C., February 15, 2012; Martin Dempsey, "Hearing to Receive Testimony on the Impacts of Sequestration and/or a Full–Year Continuing Resolution on the Department of Defense," testimony before the Senate Armed Services Committee, Washington, D.C., February 12, 2013.

[6] "EXCLUSIVE: Sen. McCain: Conditions in the Middle East 'More Dangerous Than Any Time in the Past,'" *CNN Press Room*, July 13, 2014.

[7] John Mueller, *The Remnants of War*, Ithaca, N.Y.: Cornell University, 2004, p. 1; Steven Pinker, *The Better Angels of Our Nature: Why Violence Has Declined,* New York: The Penguin Group, 2011, p. xxi.

the family, in the neighborhood, between tribes and armed factions, and among major nations and states—is all trending in the same direction: downward.[8]

An Analysis of Historical Conflict Data

To reconcile some of these different perspectives, we examined historical data on both global conflicts and terrorist activity. The aim of our effort was twofold. First, we hoped to gain a better empirical appreciation for the incidence of war and terrorist violence over the span of several decades, including when and where incidences were transpiring. Second, we hoped to glean a better understanding of what we may expect to see in the future or, at least, to identify trends in violence.

Toward these efforts, we specifically examined data from three primary sources. The first was the Armed Conflict Dataset from the Uppsala Conflict Data Program (UCDP), which includes data on both civil and interstate wars from 1946 through 2014.[9] The UCDP defines armed conflicts as "a contested incompatibility that concerns government and/or territory where the use of armed force between two parties, of which at least one is the government of a state, results in at least 25 battle-related deaths." Our second source was the Correlates of War Project, which contains data on conflict battle deaths.[10] The third primary data source was the National Consortium for the Study of Terrorism and Responses to Terrorism (START) Global Terrorism Database (GTD). The GTD is an open-source database that includes information from 1970 to 2014 on global terrorist events, defined as "the threatened or actual use of illegal force and violence by a nonstate actor to attain a political, economic, religious, or social goal through fear, coercion, or intimidation."[11]

We present here trends in armed conflict, based on the UCDP data and on trends in terrorist attacks (based on the GTD dataset). Fluctuations in each are noted. We also offer a series of maps offering a visual demonstration of where conflict has occurred at the country level. The preponderance of evidence suggests that there is support for both sides of the debate. Today, wars exclusively between states are rare indeed. However, the trends in civil conflict and civil conflicts with the involvement of other international actors present a less positive tendency. Since reaching its peak in 1991, incidence of civil wars declined dramatically until 2003. But this fall has ended, and the number of civil wars remains relatively steady at between 22 and

---

[8] For other academic voices on this topic, see Joshua S. Goldstein, *Winning the War on War: The Decline of Armed Conflict Worldwide,* New York: The Penguin Group, 2011; Peter Wallensteen, *Understanding Conflict Resolution: War, Peace and the Global System*, London: Sage, 2011.

[9] UCDP, "Battle-Related Deaths Dataset," Vol. 5., Uppsala, Sweden: Uppsala University, 2015.

[10] Meredith Reid Sarkees and Frank Wayman, *Resort to War: 1816–2007,* Washington, D.C: CQ Press, 2010.

[11] National Consortium for the Study of Terrorism and Responses to Terrorism (START), *Global Terrorism Database*, 2014.

31 conflicts. Civil wars with international involvement have also been on the rise since 2007, reaching historic highs in 2014.

The data further reveal that global terrorist attacks increased by a factor of more than 14 between 2004 and 2014. Even excluding Iraq and Afghanistan, the sharp increase in terrorist attacks remains prominent. This growth is geographically unbalanced, however. We identified a major shift in the primary locus of the violence. Before 2003, richer countries accounted for a disproportionately high number of terrorist attacks. Since then, the opposite has been true, with major growth in the number of attacks in poorer countries.

Conflict Trends: 1946 to 2014

Since its introduction in 2002, the UCDP Armed Conflict Dataset has become the predominant scholastic source of conflict analysis, especially for domestic-level conflicts. The UCDP collects information on numerous aspects of post–World War II armed violence. The dataset is also updated and revised on an annual basis. UCDP defines active conflict as the use of armed force between two parties (one of which is a government) that results in more than 25 battle deaths in a year. The dataset identifies four types of conflict: extrasystemic (colonial) conflict, interstate conflict, civil wars, and civil wars with intervention from other states. The data show an overall increase in the frequency of conflict after World War II, reaching a peak of 52 active conflicts in 1991. This formerly upward trend reversed for over a decade, dropping to just 32 conflicts in 2003. Since then, the annual number of conflicts has remained relatively stable, with 30 to 40 active conflicts per year.

Civil Wars: The Most Common and Longest Conflicts

A number of clear trends emerge when examining conflict types through time. Figure 2.1 presents the number of armed conflicts by type from 1946 to 2014. After World War II, colonial conflicts peaked at seven conflicts in 1953 and subsequently dropped steadily. There have been no colonial conflicts since 1974. Interstate conflicts are similarly infrequent, with an average of less than two per year. Over the last decade, there have been only four interstate conflicts, all of which are classified as "low intensity," defined by UCDP as less than 1,000 battle deaths per year.

Civil wars (shown in green and purple) are consistently the most common type of war since 1946, accounting for almost 75 percent of conflicts in the 1946–2014 period. Figure 2.1 reveals an upward trend in civil conflicts until the beginning of the 1990s. In 1946, there were just eight civil conflicts. By 1991, the number had peaked at 48 civil wars. The steady increase is particularly apparent in the 1970s. As the number of colonial wars dropped, active civil conflicts more than doubled.[12] However, a steep decline in civil wars began in the 1990s and continued

---

[12] Whether there is a causal relationship between the decrease in colonial wars and the increase in civil conflicts remains an open question. See Daron Acemoglu, Simon Johnson, and James A. Robinson, "Reversal of Fortune: Geography and Institutions in the Making of the Modern World Income Distribution," *Quarterly Journal of*

until 2003. It is this downward trajectory from 1991 to 2003 that John Mueller focuses on in his work.[13] Since then, the number of civil conflicts has remained relatively stable, with between 22 and 31 conflicts per year. While civil conflicts with international involvement are, historically speaking, relatively rare—with an average of 3.7 per year—there are periods with higher-than-average frequency. In 1980, there were eight active civil conflicts with international involvement. The occurrence of these conflicts has been consistently above the historic average in recent years, growing steadily since 2003 to a historic high of 13 in 2014. As a proportion of civil wars, conflicts with international involvement have tripled from 10 percent in 2003 to 33 percent in 2014. Moreover, the secondary parties in more than 60 percent of these wars include at least one permanent member of the United Nations (UN) Security Council. Major powers are not shying away from actively engaging within the civil wars of smaller nations.

Figure 2.1. The History of Conflict, 1946–2014

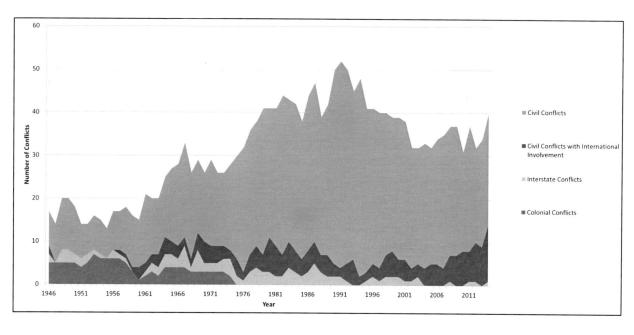

SOURCE: RAND chart based on UCDP, 2015.

Civil Conflict Duration Has Decreased, but Remains Much Longer than Interstate War

Importantly, civil conflicts tend to last significantly longer than interstate conflicts. Figure 2.2 presents the average duration of each type of conflict episode. The figure breaks up the conflicts according to the decade in which each one began. Two features of the data clearly

---

Economics, Vol. 117, No. 4, November 2002, pp. 1231–1294. On colonial legacies and civil war, see Simeon Djankov and Marta Reynal-Querol, "The Colonial Origins of Civil War," *CESifo Area Conference: Employment and Social Protection*, Munich: CESifo, May 2007.

[13] Mueller, 2004.

stand out in the figure. First, civil and colonial wars have historically lasted much longer than interstate conflicts. Second, the average duration of civil war has steadily declined since the 1970s. If, however, the many ongoing civil conflicts persist, this trend could reverse.

Figure 2.2. Conflict Duration by Decade

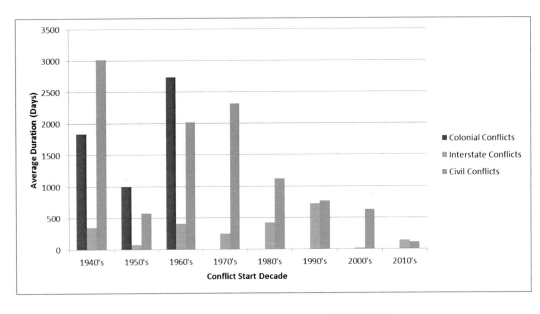

SOURCE: RAND chart based on UCDP, 2015.

In addition to duration, battle deaths provide another measure that contributes to understanding the scope and severity of each conflict. Beginning in 1989, the UCDP/PRIO battle death dataset divided conflicts between two intensity levels: low intensity (conflicts with 25 to 1,000 battle deaths per year) and high intensity (conflicts with at least 1,000 battle deaths). Figure 2.3 shows the total number of conflict-related battle deaths per year starting in 1989, sorted by civil and interstate conflicts. [14] The figure shows that, except for one period during the Eritrean-Ethiopian War, deaths from civil conflicts remain above those from interstate wars. While the average number of battle deaths per year generally falls well below the "high intensity" threshold of 1,000, civil conflicts are far more common than interstate wars. While interstate conflicts are rare, they have the potential to be extremely violent, as shown by the three respective jumps during the Gulf War, the Eritrean-Ethiopian War, and the invasion of Iraq. [15]

---

[14] UCDP, 2015.

[15] It is important to note that estimating battle deaths is challenging due to the nature of conflict and data collection issues, particularly the inclusion of noncombatant deaths. The number of deaths for a conflict can vary significantly between datasets, depending on source materials and definitions.

Figure 2.3. Annual Number of Battle Deaths by Conflict Type, 1989–2014

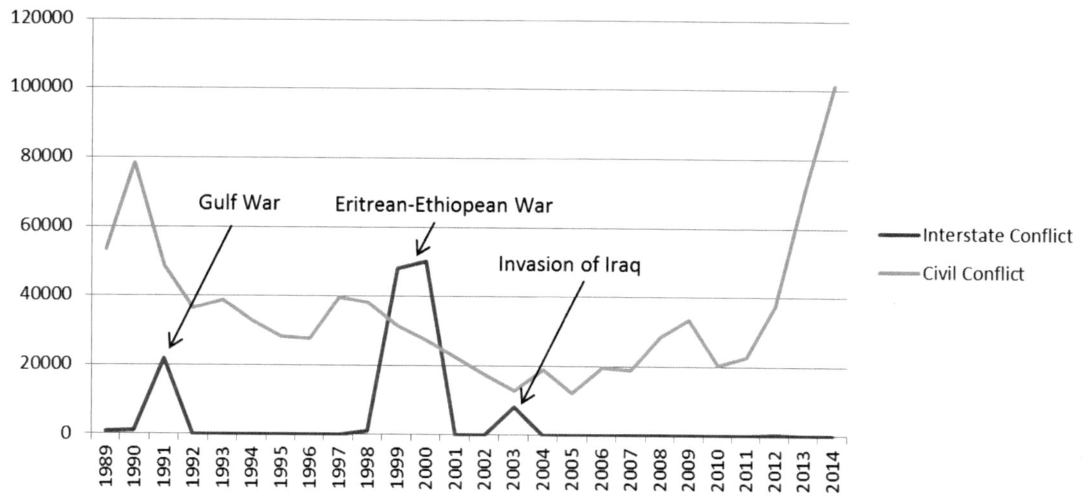

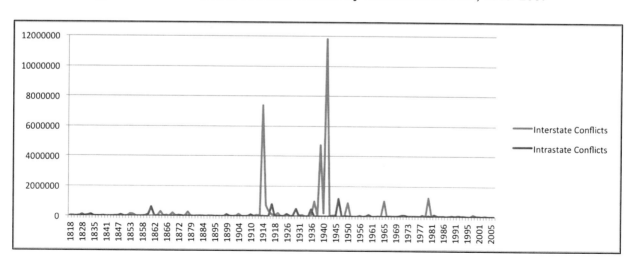

SOURCE: RAND chart based on based on UCDP, 2015.

Because the UCDP/PRIO data coverage is limited to after World War II, we supplemented the UCDP/PRIO battle death information with the Correlates of War Project, another prominent source. A review of the Correlates of War battle death data going back to 1818 revealed a similar trend. While civil conflicts tend to be more deadly on average, there are periods of particularly high violence that drive up the number of battle deaths attributed to interstate conflict. Figure 2.4 shows the relatively low number of battle deaths attributed to interstate conflicts, punctuated by large increases during World Wars I and II.

Figure 2.4. Total Number of Battle Deaths by Conflict Start Year, 1818–2007

SOURCE: RAND chart based on Correlates of War Project database.

High–Death Toll Conflicts Are Frequent in South Asia, Middle East, and Africa

The Middle East and South Asia are frequently listed as high-intensity conflict regions, with Afghanistan, Pakistan, Sri Lanka, and Iraq particularly prone to recurring conflict.[16] Figure 2.5 breaks up the number of conflicts by region. It shows that historically most of the world's conflict has been in Asia. The figure also shows that, since the mid-1970s, Africa has suffered from at least ten active conflicts, with the brief exception of 2004–2006. And while wars in the Middle East frequently capture the attention of news media in the West, there have always been more conflicts elsewhere in the world at any given time.

Figure 2.5. Number of Active Conflicts by Region: 1946–2014

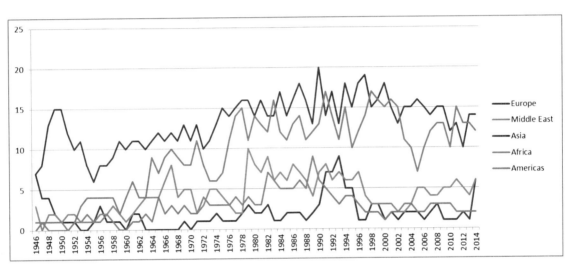

SOURCE: RAND chart based on based on UCDP, 2015.

As noted, the African continent has seen persistently high levels of conflict. Since the end of the Cold War, African countries have accounted for a large percentage of total world conflicts, generally in the 20 percent to 40 percent range. In 2014 alone, 12 African countries experienced conflicts, with six of these characterized as high-intensity wars. Figure 2.6 shows the extent of these wars, highlighting countries with both low- and high-intensity conflicts in the years 1994, 2004, and 2014. These ten-year snapshots highlight the intractability of conflict in many of these countries, particularly Uganda, Sudan, South Sudan, and Egypt.[17]

---

[16] Ongoing conflicts in Syria and Yemen are adding to this conflict toll, although post–2012 data are not available for Syria. Average civil conflict deaths in 2013 and 2014 are likely to rise when updated data become available for these conflicts.

[17] On conflict recurrence in Africa, see Barbara F. Walter, "Why Bad Governance Leads to Repeat Civil War," *Journal of Conflict Resolution,* Vol. 59, No. 7, October 2015; and David T. Mason, Mehmet Gurses, Patrick T. Brandt, and Jason Michael Quinn, "When Civil Wars Recur: Conditions for Durable Peace After Civil Wars," *International Studies Perspectives,* Vol. 12, No. 2, May 2011, pp. 171–189.

Figure 2.6. Conflict in Africa

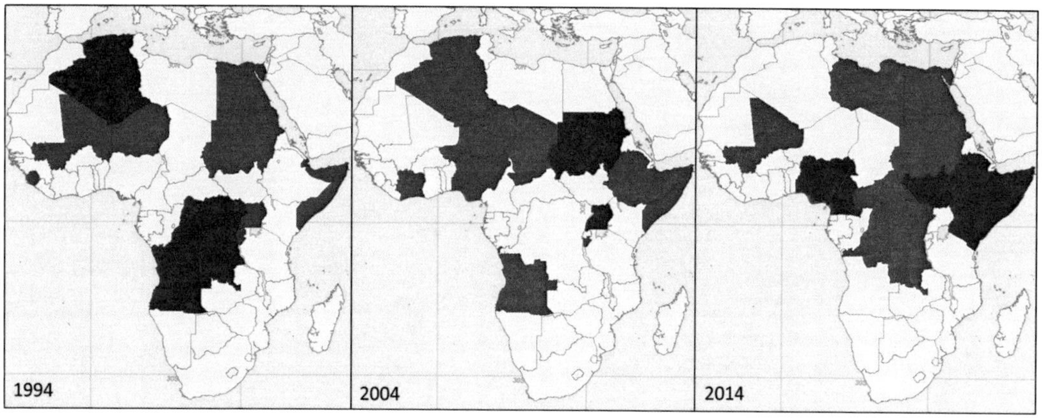

SOURCE: RAND chart based on based on UCDP, 2015.

Terrorist Attacks on the Rise Since 2004

While the UCDP dataset shows a general decline in the incidence of major state-based conflicts, the number of annual terrorist attacks has increased significantly.[18] To gain a better appreciation of this type of violence, we looked at the incidence of international terrorism. Specifically, we used the GTD to analyze the prevalence and location of terrorist attacks worldwide. The GTD defines terrorism as "the threatened or actual use of illegal force and violence by a nonstate actor to attain a political, economic, religious, or social goal through fear, coercion, or intimidation."[19] Overall, terrorist attacks are not particularly deadly. Around 50 percent of attacks in the GTD report zero fatalities. Attacks that cause fatalities average just fewer than five fatalities per attack.

Figure 2.7 displays the number of attacks worldwide by year. There is a gradual increase from the start year in 1970 until 1992, then a period of decline until 2004. For the past decade, however, there has been a dramatic rise in the number of terrorist attacks, from just over 1,000 in 2004 to almost 17,000 in 2014. The majority of these attacks are likely tied to conflict; cross-referencing with the conflict data indicates that more than 8 percent of terrorist attacks took place in countries with active wars over the past ten years. However, the upward trend of the past decade holds even if terror attacks in Iraq and Afghanistan are excluded.

---

[18] It is important to note that the methodology of the START data collection has changed as resources have become unavailable or as START staff have made efforts to improve data collection. These changes may account for some of the differences in the level of attacks. More information can be found in Michael Jensen, "Discussion Point: The Benefits and Drawbacks of Methodological Advancements in Data Collection and Coding: Insights from the Global Terrorism Database," National Consortium for the Study of Terrorism and Responses to Terrorism, November 25, 2013.

[19] Analysis of START GTD data.

Figure 2.7. Annual Terrorist Attacks

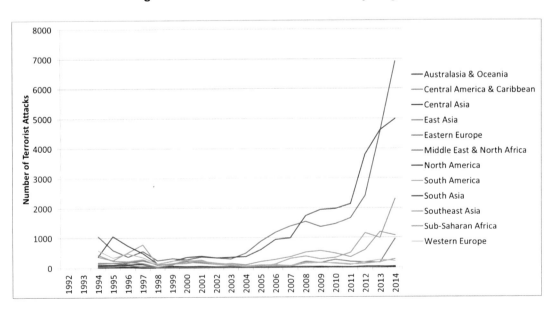

SOURCE: RAND chart based on GTD.

We may further disaggregate the data by regions. Figure 2.8 separates the number of total attacks by the regions where they occurred. As is evident in the graphic, the regions most recently and most strongly affected by terrorist attacks include South Asia, the Middle East, and North Africa. These regions account for over 70 percent of the attacks in the last ten years. Sub-Saharan Africa, however, has seen the most dramatic increase in attacks. Accounting for just 3 percent of annual attacks in 2004, the number of attacks had increased 65-fold by 2014, when 14 percent of global attacks took place in the region.

Figure 2.8. Annual Terrorist Attacks by Region

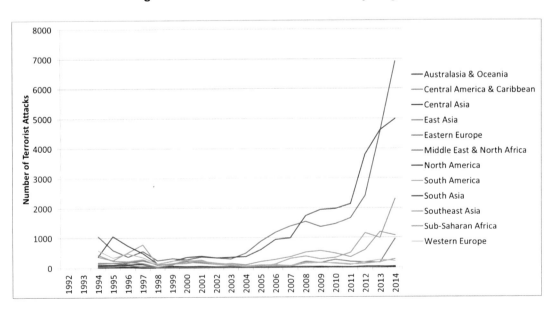

SOURCE: RAND chart based on GTD.

14

Further analysis of the data shows that the majority of terrorist attacks take place in countries that have active conflicts within their borders. This trend has increased since 1989, fluctuating in the 80–90 percent range for the last ten years. While we cannot definitively conclude that all these attacks are directly related to active conflicts, other work argues that terrorism and insurgency are often combined into the same metrics, driving up the number of reported terrorist attacks.[20] As Figure 2.9 shows, countries with active conflicts have experienced almost three times the average number of annual terrorist attacks since 2001 than in the 1989–2000 period. The average number of attacks in countries without active conflict decreased from around six to three between the same periods.

Figure 2.9. Average Number of Terrorist Attacks in Countries With and Without Active Conflicts

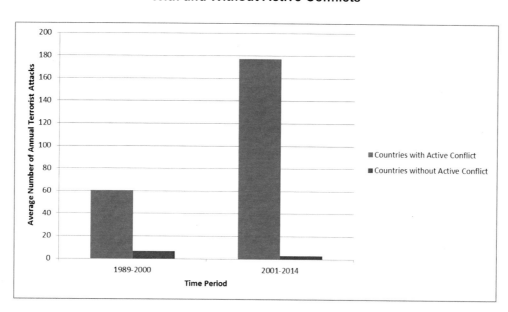

SOURCE: RAND chart based on GTD and UCDP PRIO Armed Conflict Dataset.

While these trends may appear alarming, much recent literature on terrorism challenges the conclusion that terrorists represent significant security threats.[21] In fact, research indicates that, although prolific, most terrorist organizations do not last very long and generally fail to accomplish their declared objectives. For instance, a recent report by Krusrav Gaibulloev and Todd Sandler reveals that more than a quarter of terror organizations do not survive their first year, and nearly 70 percent fail to survive for longer than ten years.[22] Other new work indicates

---

[20] John Mueller and Mark Stewart, "Conflating Terrorism and Insurgency," *Lawfare*, February 28, 2016.

[21] See Jacob Shapiro, *The Terrorist's Dilemma: Managing Violent Covert Organizations,* Princeton, N.J.: Princeton University Press, 2013.

[22] Khusrav Gaibulloev and Todd Sandler, "Determinants of the Demise of Terrorist Organizations," *Southern Economic Journal*, Vol. 79, No. 4, 2013, pp. 774–792.

that 68 percent of groups cease to meaningfully exist as a terror organization in the first year following an attack and that the average survival for any given terrorist group is a little over three years.[23]

Changing Salafi Jihadist Groups

The disaggregated regional trend lines intimate that much of the new terrorist violence may be a consequence of the changing face of Salafi jihadism. In 2008, Marc Sageman argued that the decapitation of al Qaeda leadership was leading to a dispersed network of weaker, self-organized terrorist groups.[24] Without a central organizing force, these groups are more difficult to target, but also lack the training and resources necessary to orchestrate large-scale international attacks. A 2014 report on the evolution of terrorist groups is in line with Sageman's argument, showing a significant increase in the number of Salafi jihadist groups. In the report, political scientist Seth Jones argues that this increase is the result of decentralization, showing that the number of worldwide Salafi jihadist groups doubled from 23 to 49 between 2004 and 2013.[25] The marked increase in terrorist attacks could be due to a decentralization of large groups, such as al Qaeda, and an increased focus on the "near" enemy versus the "far" enemies in the West.

The decentralization of these groups and their increasingly disparate ideologies and goals are likely seen as a greater threat by military planners, who must prepare for a multiplicity of hazards and scenarios. However, terrorism data support an increased emphasis on the "near" enemy, as outlined by Jones. Moreover, the civil wars and sociopolitical instability transpiring in regions of the world noted above may offer space and opportunity for transnational terrorist organizations to exploit. These interactive dynamics make these parts of the world less safe than other areas. When viewed as part of a larger picture, terrorism is thus far less of a threat to the Western world than it has ever been.

Terrorist Attacks Far More Likely to Affect Low-Income Populations

In fact, the pronounced increase in terrorist activity in these noted regions of the world is reflective of a broader trend in terrorist violence since the middle of the past decade. To better understand where terrorist attacks are happening, we cross-referenced terrorist attack locations with per capita gross domestic product (GDP) going back to 1989. For each year, we calculated the average per capita GDP of the countries most affected by terrorism, which we defined as

---

[23] 3.33 to be exact. Joseph K. Young and Laura Dugan, "Survival of the Fittest: Why Terrorist Groups Endure," *Perspectives on Terrorism*, 2014, Vol. 8, No. 2, pp. 1–23.

[24] Marc Sageman, *Leaderless Jihad: Terror Networks in the 21st Century,* Philadelphia: University of Pennsylvania Press, 2008, pp. 125–146.

[25] Seth G. Jones, *A Persistent Threat: The Evolution of al Qa'ida and Other Salafi Jihadists*, Santa Monica, Calif.: RAND Corporation, RR-637-OSD, 2014, p. 27.

those in the top 25 percent of terrorist attacks. We then compared this number to the global average per capita GDP. Figure 2.10 shows these averages side by side for each year.

Prior to 2004, countries that experienced the highest number of terrorist attacks tended to be wealthier, a trend largely driven by civil conflicts in the United Kingdom, Spain, and France in which domestic groups committed terrorist attacks. The figure clearly demonstrates that the countries in the top quartile of terrorist attacks (shown in blue) had a per capita GDP consistently above the global average per capita GDP (shown in red) in these years. We see a shift in this trend that coincides with a rapid increase in the number of attacks after 2004. After 2004, terrorism tended to afflict poorer states. The average per capita GDP in countries in the top quartile of terrorist attacks tends be far below the global GDP average. This dynamic is largely driven by low-income countries like Afghanistan, Kenya, and Nepal. While terrorist attacks have increased throughout the world, Western Europe and North America are among the regions with the lowest growth in terrorist activity, with only 2.6–percent and 1.2–percent growth in the last ten years, respectively.[26] Worryingly, these poorer states are precisely the countries with higher risks of experiencing civil conflicts. The confluence of terrorism and civil war is not by chance.

Figure 2.10. Per Capita GDP in Countries with High Terrorist Attack Frequency vs. Global Averages

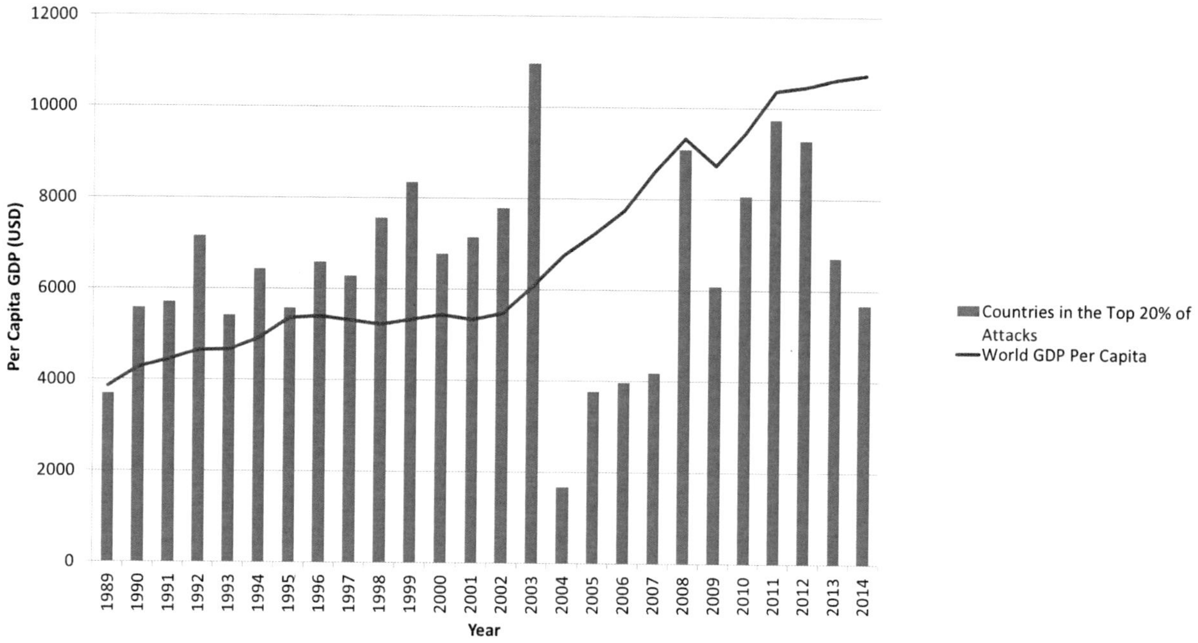

SOURCE: RAND chart based on GTD and the "GDP per Capita (current US$)," web page, n.d.

---

[26] Analysis of START GTD data.

While the global upward trend in terrorist attacks and Salafi jihadist violence is striking, it is important to remember that, on an annualized basis, terrorism accounts for very few deaths worldwide and even fewer within the United States.[27] For example, in 2013, 112 Americans died in terrorist attacks.[28] In contrast, 12,253 homicides occurred nationwide,[29] and 32,719 Americans died that same year in motor vehicle accidents.[30]

Conclusions

As the broader conflict data of the past 70 years has shown, trends are not necessarily long lasting and often conceal significant changes in the distinct aspects of conflict, such as the dramatic increase in terrorist attacks. Furthermore, it is often difficult to distinguish between terrorism and insurgency, especially for methodological purposes. Some examples help illustrate the point. ISIS is an insurgent group, but it also conducts acts of terrorism outside its theater of operations. The Afghan Taliban is also an insurgent group that uses terrorist tactics in theater but is not a terrorist organization like al Qaeda. The Pakistani Taliban may more accurately be described as both an insurgent and a terrorist group. Al-Shabaab in Somalia originally was an insurgency with a nationalist agenda that has only in recent years turned its attention to terrorism in Uganda and Kenya because of the actions those countries have undertaken in Somalia. Similarly, Boko Haram represents an insurgent group that often resorts to terrorism. These examples help account for some of the trends described above. They also demonstrate that separating terrorism from conflict may be increasingly challenging. In light of such nuances, predicting conflicts 30 years into the future will be particularly difficult, and linking such trends to posture demands is likely not the most propitious approach for achieving a nimble, responsive posture.

The purpose of this brief review is not to try to tie posture needs to incidents of conflict. Rather, the summary of intra- and interstate conflict and terrorist violence helps put into perspective a topic that is often misunderstood or misconstrued. Interstate conflicts are no longer as common as they use to be; indeed, they are exceedingly rare. Civil wars are still prevalent, but not at the same frequency as during the end of the Cold War and the years immediately after.[31] Although reported incidents of terrorist violence are at all-time highs, the evidence examined

---

[27] Navin A. Bapat, "The Escalation of Terrorism: Microlevel Violence and Interstate Conflict," *International Interactions,* Vol. 40, No. 4, 2014, pp. 568–578.

[28] Analysis of START GTD data.

[29] Federal Bureau of Investigation, *Crime in the United States: 2013*, Washington, D.C., 2013.

[30] National Highway Traffic Safety Administration, "U.S. Department of Transportation Announces Decline in Traffic Fatalities in 2013," December 19, 2014.

[31] Our findings are broadly in line with new work on conflict trends. For a good and relatively recent review of this debate, see Nils Petter Gleditsch, Steven Pinker, Bradley A. Thayer, Jack S. Levy, and William R. Thompson, "The Forum: The Decline of War," *International Studies Review,* Vol. 15, No. 3, 2013, pp. 396–419.

here suggests that the overwhelming majority of these attacks occur in countries with extant civil conflicts. This finding is in line with work in conflict studies demonstrating that weak states are hubs of terrorist activity.[32]

The key issue for USAF is the risk that future wars and terrorist activity pose to U.S. interests—the scale of which is difficult to anticipate. Conflict with a major power would likely place much greater demands on USAF posture than terrorist activity, but such conflicts are relatively rare. However, the current proportion of civil conflicts with international involvement is higher than it has ever been in the post–World War II period. Moreover, terrorist attacks are more frequent and rising in poorer, less stable regions of the world—precisely where civil conflicts are more likely. Such civil conflicts appear to be contributing to the spread of terrorist activity. It may therefore be the interplay between these phenomena that proves most dangerous. Transnational terrorist groups take advantage of civil wars, often finding safe havens within the borders of relatively unstable states. This may provoke the involvement of neighboring countries and other states in the international system. The interaction of the various conflict events chronicled above could present the greatest challenge to USAF. On top of this, regional trends suggest that this dynamic may continue to play out in the Middle East, North Africa, and Central Asia. Preserving the capability to respond or take military action in these regions will be an important consideration for the foreseeable future.

The relationship between long-term conflict trends and USAF posture is complex. As will be discussed in Chapter Five, the most common driver of big posture changes is a conflict or crisis (e.g., World War II). Once basing is acquired, path dependent processes often result in prolonged stays whether or not there are ongoing conflicts in the region. Since U.S. strategy emphasized conflict deterrence for most of the Cold War, USAF posture was heavily influenced by deterrence requirements as opposed to trends in past or current conflicts. Alternatively, ongoing conflicts may be in locations or regions that are not deemed to be of sufficient interest to the United States.

Whether the overall security environment becomes more or less threatening may actually be a secondary concern compared to the eclectic nature of emerging threats and their frequency, scale, and type. In many ways, the world is becoming less violent, but this does not automatically translate into a reduction in demands for U.S. military—specifically USAF—capabilities, capacity, or readiness. Conflict trends offer insights into the potential location and type of demands on USAF (and its posture) over the near term, but the volatility in trends over longer periods limits their utility in projecting USAF posture needs over the 30-year planning horizon.

---

[32] See James A. Piazza, "Incubators of Terror: Do Failed and Failing States Promote Transnational Terrorism?" *International Studies Quarterly,* Vol. 52, No. 3, 2008, pp. 469–488.

3. Strategy Choices and Posture Implications

This chapter explores whether and how grand and military strategies drive the shape, size, and character of the U.S. overseas military presence. (For our purposes, "strategy is the art of controlling and utilizing the resources of a nation—or a coalition of nations—including its armed forces, to the end that its vital interests shall be effectively promoted and secured against enemies, actual, potential or merely presumed."[33]) We begin with a discussion of grand strategy—specifically, how the level of overseas activity and type of security commitment changed as the United States moved from a grand strategy of isolationism (at one extreme) to our current grand strategy of deep engagement. The chapter then considers the relationship between U.S. military strategy and global defense posture as they have evolved since the end of World War II.

In the United States, deliberations about the U.S. overseas military presence center on how DoD needs to alter its posture to better meet emerging threats in a period of fiscal austerity. For instance, Michael O'Hanlon and Bruce Riedel argue that the United States should permanently station a large force of land-based fighter aircraft in the Persian Gulf to reduce the nation's dependence on expensive aircraft carriers. Others have challenged this proposal on the grounds that it restricts USAF flexibility by tying down a large number of combat aircraft and increases the probability that DoD will encounter access problems with its partners in the Gulf. According to this view, a mixed presence of ground- and sea-based combat aircraft is the appropriate posture.[34]

Similarly, China's rapidly expanding military capabilities and its more assertive behavior provided a major rationale for DoD to announce that it was pivoting (or rebalancing) toward the Asia-Pacific region.[35] As a part of this initiative, DoD is seeking additional access in Southeast Asia and Oceania to improve the resiliency of its posture. Debates surrounding this policy have centered on whether the components of the pivot represent a significant change to the current posture; the sustainability of this shift; its effect on other critical regions, such as the Middle East; and the risks of provoking an insecure China.[36] Most recently, in response to ongoing the

---

[33] Edward Mead Earle, quoted in Paul Kennedy, *Grand Strategies in War and Peace*, New Haven, Conn.: Yale University Press, 1992, p. 2.

[34] They specifically call for 150 aircraft. See Michael O'Hanlon and Bruce Riedel, "Land Warriors: Why the United States Should Open More Bases in the Middle East," *Foreign Affairs,* July 2, 2013; Stacie L. Pettyjohn and Evan Braden Montgomery, "By Land and by Sea," *Foreign Affairs,* July 22, 2013.

[35] Hillary Clinton, "America's Pacific Century," *Foreign Policy*, October 11, 2011; and DoD, *Sustaining U.S. Global Leadership: Priorities for 21st Century Defense*, Washington, D.C., January 2012.

[36] Mark E. Manyin, Stephen Daggett, Ben Dolven, Susan V. Lawrence, Michael F. Martin, Ronald O'Rourke, and Bruce Vaughn, *Pivot to the Pacific? The Obama Administration's 'Rebalancing' Toward Asia*, Washington, D.C.: Congressional Research Service, March 28, 2012; Robert G. Sutter, Michael E. Brown, and Timothy J. A. Adamson,

Russo-Ukrainian crisis, there have been calls for the United States to expand its military presence in Eastern and Central Europe to deter further Russian aggression.[37] All these examples presume that the U.S. posture is largely driven by strategic considerations and that it can easily be expanded, contracted, or realigned as needed. If basing decisions were driven solely by strategy, USAF would simply select the course of action—whether opening a new base, realigning bases, or closing a base—that is expected to best achieve its objective of enhancing U.S. security. In other words, basing decisions would be made through a rational process that aligned ends, means, and costs. Although real-world decisions are not quite this simple, at the least, strategy plays an important role in framing and articulating policy choices.

Two primary types of strategies—grand and military—can influence posture. There is little consensus on what actually constitutes grand strategy; we define it as a logic—or "an integrated conception of interests, threats, resources, and policies"—that guides a nation's actions with the world.[38] Grand strategy is rooted in perceptions of a nation's enduring interests and offers a plan for how to achieve its goals. As the overarching rubric within which defense and foreign policy is made, a nation's grand strategy will obviously shape, but not determine, its military strategy. Given that grand strategies are fairly abstract theories about interests, threats, and resources, there may be numerous different military strategies that would be congruent with a particular grand strategy.

Grand Strategy

The United States has maintained the same grand strategy—deep engagement—for over 70 years. Nevertheless, future leaders may decide that deep engagement is no longer an appropriate or sustainable strategy given changing international circumstances or dwindling resources. Moreover, different grand strategies have very different implications for U.S. overseas military presence. Figure 3.1 depicts four ideal types of grand strategies that fall along a

---

Balancing Acts: The U.S. Rebalance and Asia-Pacific Stability, Washington, D.C.: Elliott School of International Affairs, George Washington University, August 2013, pp. 26–28; Ely Ratner, *Resident Power: Building a Politically Sustainable U.S. Military Presence in Southeast Asia and Australia,* Washington, D.C.: Center for New American Security, October 2013; Bilal Y. Saab, "Asia Pivot, Step One: Ease Gulf Worries," *National Interest,* June 20, 2013; and Robert S. Ross, "The Problem with the Pivot: Obama's New Asia Policy Is Unnecessary and Counterproductive," *Foreign Affairs,* November/December 2012.

[37] Andrew A. Michta, "U.S. Needs New Bases in Central Europe," *American Interest,* June 11, 2014; Terrence Kelly, "Stop Putin's Next Invasion Before It Starts," *U.S. News and World Report,* March 20, 2015. For what the United States is doing, see U.S. European Command, "EUCOM Provides Update on the European Reassurance Initiative," April 20, 2015; Eric Schmitt and Steven Lee Myers, "U.S. Is Poised to Put Heavy Weaponry in Eastern Europe," *New York Times,* June 13, 2015.

[38] Hal Brands, *The Promise and Pitfalls of Grand Strategy,* Carlisle, Pa.: U.S. Army War College, Strategic Studies Institute, August 2012, p. 3-4. Other definitions include "A grand strategy is a nation-state's theory about how to produce security for itself," from Barry R. Posen, *Restraint: A New Foundation for U.S. Grand Strategy,* Ithaca, N.Y.: Cornell University Press, 2014, p. 1. According to Robert J. Art, "a grand strategy tells a nation's leaders what goals they should aim for and how best they can use their country's military power to attain these goals" (Robert J. Art, *A Grand Strategy for America,* Ithaca, N.Y.: Cornell University Press, 2003, p. 1).

spectrum of how engaged the United States is abroad. Moving from the least amount of overseas involvement to the most, they are isolationism, offshore balancing, selective engagement, and deep engagement. This is not an exhaustive list of possible U.S. grand strategies. It excludes those that focus on specific means (such as collective security) and focuses on strategies that offer the starkest choices for posture.

Figure 3.1. Grand Strategies and Involvement Overseas

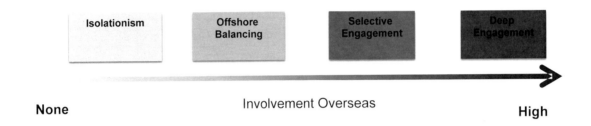

Isolationism, or the desire to avoid overseas entanglements, has deep roots in the United States.[39] In his farewell address, President George Washington warned the nation to "steer clear of permanent alliances" that risked entrapping the United States in costly conflicts.[40] In part, isolationism is an option for the United States because it is a "geographic fact"—no other great powers exist in the Western Hemisphere.[41] Isolationists have a narrow conception of U.S. interests, centering on homeland defense. In their view, the United States is an inherently secure nation protected by the Atlantic and Pacific Oceans, while much of the rest of the world is plagued with instability. Therefore, the best option for the United States is to remain neutral and stay out of the fray. Because there are no real threats to justify the presence of U.S. forces abroad, isolationists favor the complete retrenchment of U.S. forces and the severing of all U.S. overseas security commitments. Today, there are few advocates of strict isolationism, and this seems like the grand strategy that the United States is least likely to adopt—at least in the near term. However, there are a growing number of academic proponents of offshore balancing, which is a less strict version of isolationism.

Offshore balancing is a strategy modeled after the United Kingdom's traditional practice of remaining on the sidelines of continental European conflicts and intervening only if one nation was going to dominate the region.[42] Like isolationists, offshore balancers believe that geography, nuclear weapons, and conventional military superiority combine to make the United States quite

[39] Walter A. McDougall, *Promised Land, Crusader State: The American Encounter with the World Since 1776*, New York: Houghton Mifflin Company, 1997, pp. 39–56.

[40] George Washington, "Washington's Farewell Address," 1796.

[41] McDougall, 1997, p. 40.

[42] John J. Mearsheimer, *The Tragedy of Great Power Politics*, New York: W.W. Norton, 2001, pp. 261–264.

safe.[43] The only near-term threat that the United States faces is the prospect of nuclear war, which is heightened by maintaining extended deterrent commitments to other nations; therefore, the United States should dissolve its security alliances. Ending security alliances also would have the benefit of ending the pervasive practice of freeriding, forcing America's allies to provide for their own security. Like isolationists, offshore balancers believe that U.S. military activity overseas is costly and counterproductive, provoking anti-American sentiment and terrorism. Remaining overseas tempts U.S. leaders to meddle in the affairs of others and inevitably will lead to overextension and decline. They advocate that the United States should pull most of its forces back and rely primarily on maritime forces to ensure the freedom of the commons. Aside from securing the seas, U.S. forces should intervene in a conflict only if it appears as if one great power is going to dominate a majority of the others.

In contrast, *selective engagement* proponents hold that the U.S. military needs to remain moderately involved overseas to ensure national security.[44] At the same time, proponents of this view argue that the United States has gone far beyond protecting its vital interests and has intervened in the domestic affairs of other states that have little bearing on U.S. security. Doing so has sapped U.S. resources and involved it in conflicts that are difficult—if not impossible—to win. Selective engagers believe that the only real threat to U.S. security is great-power war. Consequently, the United States should maintain its security commitments overseas but refrain from intervening in civil or internal conflicts, such as in the Balkans. In short, the United States remains the security guarantor for close partners to ensure stability but stays out of messy and costly internal conflicts and is "sparing in its use of the military instrument."[45] Selective engagers largely support the preservation of existing alliances and posture but believe that the United States needs to be more disciplined and discriminating in when and how it utilizes military power.

---

[43] Posen, 2014; Barry R. Posen, "The Case for Restraint," *American Interest*, Vol. 3, No. 1, November/December 2007, pp. 7–17; Stephen M. Walt, *Taming American Power: The Global Response to U.S. Primacy*, New York: W.W. Norton, 2006; Eugene Gholz, Daryl G. Press, and Harvey M. Sapolsky, "Come Home, American: The Strategy of Restraint in the Face of Temptation," *International Security*, Vol. 21, No. 4, Spring 1997, pp. 5–48; Paul K. MacDonald and Joseph M. Parent, "Graceful Decline? The Surprising Success of Great Power Retrenchment," *International Security*, Vol. 35, No. 4, Spring 2011, p. 7–44; Christopher A. Preble, *Power Problem: How American Military Dominance Makes Us Less Safe, Less Prosperous and Less Free*, Ithaca, N.Y.: Cornell University Press, 2009; Christopher Layne, *The Peace of Illusions: American Grand Strategy from 1940 to the Present*, Ithaca, N.Y.: Cornell University Press, 2006; Christopher Layne, "From Preponderance to Offshore Balancing: America's Future Grand Strategy," *International Security*, Vol. 22, No. 1, Summer 1997, pp. 86–124; Robert A. Pape, "Empire Falls," *National Interest*, No. 99, January/February 2009, pp. 21–34; and Robert A. Pape, *Dying to Win: The Strategic Logic of Suicide Terrorism*, New York: Random House, 2005.

[44] Art, 2003; Barry Posen and Andrew L. Ross, "Competing Visions for Grand Strategy," *International Security*, Vol. 21, No. 3, Winter 1996–1997.

[45] Art, 2003, p. 11.

Finally, there is *deep engagement*, which is the current American grand strategy.[46] Deep engagers believe U.S. leadership is necessary to create a stable, peaceful, and prosperous world. The objectives of this strategy are to shape the external environment to prevent the emergence of short- and long-term threats, maintain a liberal global economic order, and sustain institutionalized cooperation. These goals rest upon two pillars: maintaining U.S. security commitments and taking a leading role in global affairs. The United States does not assume this role for altruistic reasons; this strategy disproportionately benefits the United States and helps to preserve its hegemony. Moreover, in the absence of U.S. involvement, the world would be a much more dangerous place. Deep engagers thus believe that it is not only less costly but also beneficial for the United States to remain deeply involved in international affairs to shape the environment and prevent threats from developing.

Today, there is a strong consensus within the U.S. policy community in favor of deep engagement.[47] But initially, this was not the case. During and immediately after World War II, when the Joint Chiefs of Staff (JCS) advocated for a more activist and internationalist U.S. strategy that included a network of U.S. overseas bases, they encountered significant resistance at home and abroad. The JCS had not identified any potential adversary yet felt that the United States needed access to a network of foreign air and naval bases so that threats could "be met [as] far from our borders as possible."[48] The JCS sought to create a network of overseas bases along the perimeters of the European and Asian continents that would allow the United States "to rapidly deploy forces in any desired direction."[49] This posture of perimeter defense-in-depth emphasized offshore locations that could strike in many directions and using basing rights instead of large fixed garrisons, which would help adaptation to changing circumstances. The JCS therefore wanted to develop a posture that would allow the United States to respond to threats wherever they might emerge and that rested on the United States being deeply engaged with the rest of the world.[50] However, few initially supported this strategy in the United States or overseas. At home, the U.S. military was demobilizing, and a lack of sufficient resources forced the JCS to scale back its initial basing plans. Additionally, other nations were reluctant to allow

---

[46] Stephen G. Brooks, G. John Ikenberry, and William C. Wohlforth, "Don't Come Home, America: The Case Against Retrenchment," *International Security*, Vol. 37, No. 3, Winter 2012–2013, pp. 7–51; Stephen G. Brooks, G. John Ikenberry, and William C. Wohlforth, "Lean Forward: In Defense of American Engagement," *Foreign Affairs*, January/February 2013.

[47] Posen, 2014, pp. 5–10.

[48] JCS, "United States Military Requirements for Air Bases, Facilities, and Operating Rights in Foreign Territories," JCS 570/2, Reference Group 218, Combined Chief of Staff series 360 (12-9-42), November 2, 1943.

[49] JCS, "Enclosure C: Overall Examination of United States Requirements for Military Bases and Base Rights," JCS 570/40, RG 218, Combined Chiefs of Staff series 360 (19-9-42), November 7, 1945.

[50] Elliot V. Converse III, *Circling the Earth: United States Military Plans for a Postwar Overseas Military System, 1942–1948*, Maxwell Air Force Base, Ala.: Air University Press, 2005, p. 132.

U.S. forces to be indefinitely stationed on their territories during peacetime.[51] As a result, the JCS plans were never fully implemented, although the notion that the United States needed to defend forward was internalized by all successive generations of U.S. leaders.

When the Cold War began in the late 1940s, U.S. political and military officials desired access to foreign bases for U.S. long-range bombers to deter Soviet aggression. The outbreak of the Korean War sparked widespread fears that communism, left unchecked, would continuously expand. In this situation, a network of relatively austere air and naval bases was seen as insufficient. Because U.S. allies were too weak to defend themselves, they encouraged the United States to station large numbers of ground, air, and naval forces in critical battleground areas, especially the European Central Front and Northeast Asia. As a result, the United States created an expansive network of overseas garrisons placed to defend fixed locations. This network differed significantly from the early JCS plans but played an integral part in containment strategy.[52]

While there is a lively academic debate over U.S. grand strategy, there is little disagreement in the policy realm that deep engagement is the appropriate strategy.[53] Nevertheless, in the time span of decades, the United States quite possibly could modify its grand strategy, dramatically affecting the shape and size of its overseas military presence.

Military Strategy

Although U.S. officials often have the same overarching grand strategy, they may hold different views on how the military can best achieve these ends. Consequently, an additional strategic factor that influences basing decisions is the administration's preferred defense strategy—that is, "the link between military and political ends, the scheme for how to make one produce the other."[54] Because a defense strategy may prioritize certain threats, military capabilities, or specific regions, the particular strategy adopted by an administration affects the shape of USAF posture.[55]

During the Cold War, for instance, all U.S. presidents agreed that the Soviet Union posed a threat to the United States and therefore had to be contained. Yet the Truman, Eisenhower, and Kennedy administrations differed in how they used U.S. forces to deter Soviet aggression. After

---

[51] Melvyn P. Leffler, "The American Conception of National Security and the Beginnings of the Cold War, 1945–48," *The American Historical Review*, Vol. 89, No. 2, April 1984, p. 352.

[52] For the creation of this posture, see Stacie L. Pettyjohn, *U.S. Global Defense Posture, 1783–2011*, Santa Monica, Calif.: RAND Corporation, MG-1244-AF, 2012, pp. 61–66.

[53] Posen, 2014, p. 5.

[54] Richard K. Betts, "Is Strategy an Illusion?" *International Security*, Vol. 25, No. 2, 2000, p. 5.

[55] Some readers may be surprised that we do not discuss the role of operational plans as drivers of posture. We argue that they have a relatively small effect on USAF global posture. See Pettyjohn and Vick, 2013, pp. 70–72, for details.

facing Soviet-backed communist aggression on the Korean Peninsula, the Truman administration adopted an expansive strategy, articulated in National Security Council (NSC)-68, of countering communist aggression across the globe.[56] This included rolling back communist expansion in Korea and deterring the outbreak of a general war in Europe. The adoption of NSC-68 led to a dramatic expansion of USAF's network of bases at home and abroad. Because of the magnitude of the threat that the United States faced, NSC-68 called for comprehensive rearmament, avoiding prioritizing one mission or subset of missions over the others. For USAF, this meant that, while the strategic nuclear mission remained the foundation of the U.S. deterrent, nuclear weapons alone were not believed to be capable of winning a protracted war against the Soviet Union.[57] Instead, the Truman administration believed that the United States also needed formidable conventional forces, including large numbers of tactical aircraft stationed abroad.[58] In Europe, for example, the United States went from 14 air bases (ten of which were standby or support bases) used primarily for occupation duties in 1947 to 42 major air bases by 1953.[59]

President Dwight D. Eisenhower questioned the assumption that underlay the Truman administration's vast rearmament program: that resources would expand to meet security requirements. Consequently, the Eisenhower administration's New Look strategy stressed that containment needed to be affordable because a strong free market economy was the foundation of U.S. military strength.[60] Eisenhower feared that unrestrained defense spending would undermine the U.S. economy and maintained that the United States needed "security with solvency."[61] The New Look strategy, therefore, emphasized the deterrent value of nuclear weapons and continental defenses, while deemphasizing conventional forces.[62] According to this logic, deterrence rested primarily on the credible threat to massively retaliate against any aggression with nuclear weapons, coupled with strengthened air defenses. Because of this focus on nuclear retaliation, Strategic Air Command (SAC) and, to a lesser extent, Air Defense Command received the preponderance of USAF's budget and base structure. To reduce military expenditures while expanding nuclear and air defense forces, the Eisenhower administration cut

---

[56] National Security Council, "United States Objectives and Programs for National Security," Washington, D.C., April 12, 1950.

[57] Allan R. Millett and Peter Maslowski, *For the Common Defense: A Military History of the United States of America*, New York: The Free Press, 1994, p. 516.

[58] George F. Lemmer, "Bases," in Alfred Goldberg, ed., *A History of the United States Air Force*, New York: Arno Press, 1974, p. 141; Millett and Maslowski, 1994, pp. 517–518.

[59] Lawrence Benson, *USAF Aircraft Basing in Europe, North Africa, and the Middle East, 1945–1980*, Ramstein Air Base, Germany: Headquarters, U.S. Air Forces in Europe, 1981, Declassified on July 20, 2011, p. 11; Pettyjohn and Vick, 2013, p. 67.

[60] John Lewis Gaddis, *Strategies of Containment: A Critical Appraisal of American National Security Policy During the Cold War*, New York: Oxford University Press, 2005, pp. 130–132; Samuel P. Huntington, *The Common Defense: Strategic Programs in National Politics*, New York: Columbia University Press, 1961, pp. 64–84.

[61] Quoted in Millett and Maslowski, 1994, p. 534.

[62] Huntington, 1961, p. 78.

the budget of other USAF commands, also leading to reductions in their base structures.[63] Despite the fact that SAC was developing new operating concepts that relied less on vulnerable forward bases, many new overseas bases were still built to support operations, while plans for several U.S. Air Forces in Europe (USAFE) bases for tactical forces were canceled.[64]

This changed when John F. Kennedy entered the White House in 1961. Kennedy and Secretary of Defense Robert McNamara believed that there were many circumstances in which it was not appropriate to employ nuclear forces and that, therefore, the Eisenhower administration's threat of massive retaliation lacked credibility.[65] Believing that a strong deterrent rested on the threat of a calibrated, proportional response, the Kennedy administration shifted to a defense strategy of flexible response.[66] Flexible response's emphasis on conventional forces was reflected in budget and infrastructure allocation among USAF commands.[67]

In Europe, for example, the ratio of SAC to USAFE bases shifted in the favor of the latter, reflecting DoD's renewed emphasis on supporting ground forces.[68] By 1960, however, there were increasing pressures on USAF's overseas military presence.[69] In 1963, DoD tried to stem the growing U.S. balance of payments deficit by mandating that USAF cut its overseas base structure. While there were cuts across the board, SAC bore the brunt of these reductions, and by 1965, SAC had vacated its bases in Europe. In the 1960s, U.S. forces were also expelled from France, Morocco, and Libya. As a result of these fiscal and political constraints, USAFE found that it did not have sufficient air bases to fight and win a conventional war in Europe.[70] USAF base structure could not support the strategy of flexible response.[71]

---

[63] Daniel L. Haulman, "Air Force Bases, 1947–1960," in Frederick J. Shaw, ed., *Locating Air Force Base Sites: History's Legacy*, Washington, D.C.: Air Force History and Museums Program, U.S. Air Force, 2004, p. 75.

[64] Benson, 1981. For more on SAC's concept of operations, see: History and Research Division Directorate of Information, *Overseas Bases: A Military and Political Evaluation*, April 2, 1962.

[65] Lawrence S. Kaplan, Ronald D. Landa, and Edward J. Drea, *History of the Office of the Secretary of Defense, Vol. V: The McNamara Ascendancy, 1961–1965*, Washington, D.C.: Historical Office, Office of the Secretary of Defense, 2006, pp. 293–294.

[66] Gaddis, 2005, pp. 214–215; Millett and Maslowski, 1994, p. 553.

[67] Between 1961 and 1969, the number of major SAC bases in the United States fell from 46 to 28 (Forrest L. Marion, "Retrenchment, Consolidation, and Stabilization, 1961–1987," in Frederick J. Shaw, ed., *Locating Air Force Base Sites: History's Legacy*, Washington, D.C.: Air Force History and Museums Program, U.S. Air Force, 2004, p. 107).

[68] The number of SAC bases had already been reduced due to changing operational concepts. SAC realized that its forward-based forces were vulnerable and, therefore, retained only bases needed for poststrike support (Benson, 1981, pp. 32, 55).

[69] Benson, 1981, p. 21. SAC was able to abandon its European bases because of changes to its force structure. In particular, SAC accelerated the retirement of its medium-range B-47 bomber, which needed to be forward based to be in range of the USSR, and increasingly relied on the intercontinental B-52 bomber.

[70] At the time, the North African air bases were a part of USAFE.

[71] Benson, 1981, p. 22.

After the Cold War, U.S. defense strategy continued to evolve. For instance, the George W. Bush administration tried to transform the U.S. military so that it would be better able to deal with unpredictable and unconventional threats. As a part of this initiative, the Bush administration's 2004 Global Defense Posture Review (GDPR) sought to enable the United States to deal with an extremely fluid and uncertain environment characterized by asymmetric threats. Because the Bush administration believed that the "principal characteristic of security environment" was "uncertainty," the 2004 GDPR proposed a number of changes to improve the agility of U.S. armed forces.[72] In particular, the GDPR sought to eliminate static formations in Europe and Asia by moving away from large main operating bases (MOBs) in favor of access to cold and warm facilities that could be scaled up or scaled down as needed.[73] A nimble posture consisting primarily of expandable facilities would allow U.S. forces "to reach any potential crisis spot quickly."[74] This initiative involved fewer base closures for USAF, which had already drawn down its posture in Europe and Asia, than for the Army. Instead, USAF found itself seeking access to new facilities in Southern and Eastern Europe, Central Asia, and Africa.

The 9/11 attacks played an important role in reducing resistance to the GDPR by emphasizing what the Bush administration suspected—DoD Cold War planning assumptions were less relevant in this new security environment. Moreover, al Qaeda's attacks on the United States demonstrated that new challenges posed an immediate and serious threat to the nation and helped convince other nations to provide access to the United States. In turn, this helped convince DoD bureaucracies that they needed to adapt to deal with these adversaries. It also galvanized U.S. legislators, who dropped their normal objections to building bases overseas and resourced the efforts to develop a network of new facilities to counter terrorism.

Nevertheless, implementing such a sweeping transformation of the U.S. global defense posture proved to be difficult. While DoD had a coherent and truly global vision of the posture that it wanted, the changes had to be negotiated bilaterally with each host nation and implemented separately, increasing the opportunities for opponents to scuttle the planned changes. Unsurprisingly, this was a drawn-out process with many setbacks and complications. For instance, the decision to reduce the U.S. Army presence in Germany was not completed until 2013, when the last U.S. heavy forces left Europe. But less than a year later, as a measure to strengthen deterrence against Russia, the Army returned to Germany some of the same armored vehicles that had been shipped home the previous year. This equipment was prepositioned in

---

[72] JCS, *The National Military Strategy of the United States of America: A Strategy for Today; a Vision for Tomorrow*, Washington, D.C.: Office of the Chairman of the Joint Chiefs of Staff, 2004, p. 7; and DoD, *Strengthening U.S. Global Defense Posture Report to Congress*, Washington, D.C.: September 2004, pp. 9–15.

[73] Ryan Henry, "Transforming the U.S. Global Defense Posture," in Carnes Lord, ed., *Reposturing the Force: U.S. Overseas Presence in the Twenty-First Century*, Newport, R.I.: Naval War College Press, 2006, p. 38; DoD, 2004, p. 10.

[74] Henry, 2006, p. 39.

Germany as a part of the U.S. Army European Activity Set.[75] Similarly, the realignment of U.S. forces in Korea and in Japan has fallen far behind their original schedules. To gain new basing rights, George W. Bush's administration often had to offer side payments, which tended to create volatile and problematic relationships.[76] For instance, to support the war in Afghanistan, the United States needed air bases in Central Asia—where DoD had little presence before 9/11. Using a variety of inducements, including arms sales and rental payments, the Bush administration was able to secure access to former Soviet air bases in Uzbekistan and Kyrgyzstan. Yet the U.S. military presence proved to be controversial, and ultimately, both nations expelled U.S. forces.

The next major change in defense strategy occurred in 2012, when President Barack Obama's administration declared that DoD would begin a process of strategic rebalancing, "pivoting" its attention and resources from the Middle East to the Asia-Pacific.[77] The logic behind this new strategy was sensible. Not only is the Asia-Pacific one of the most economically dynamic regions in the world, it is also in a delicate state: China's economic and military power continues to grow; it appears more willing to press its claims to disputed territories in the East and South China Seas; and U.S. allies with deep economic ties to China are hoping to avoid being caught between Washington and Beijing.

The pivot is supposed to help manage these issues. By shoring up America's military position in the Western Pacific, Washington should be better positioned to discourage hostile behavior by China, reassure local allies that it is willing and able to deter any attempts at aggression and coercion, and preserve stability throughout the region. So far, however, the steps that have constituted the rebalance have been relatively modest, including plans to shift additional naval forces from the Atlantic to the Pacific; station several littoral combat ships, the Navy's newest vessels, in Singapore; deploy up to 2,500 Marines in northern Australia; and increase the frequency of U.S. military deployments to the Philippines.[78] Greater access to Philippine bases was delayed because of a constitutional challenge to the U.S.–Philippine defense agreement, which was finally upheld by the Philippine Supreme Court in January 2016.[79] Similarly, expanding the U.S. military presence in Australia has proceeded more slowly than U.S. officials had hoped. While the buildup to deployment of 2,500 Marines in northern Australia has only been slightly pushed back to 2017, the U.S. and Australian governments have

---

[75] John Vandiver, "US Army's Last Tanks Depart from Germany," *Stars and Stripes,* April 4, 2013; U.S. Army Europe, "U.S. Army European Activity Set," n.d.

[76] For more on transactional relationships, see Pettyjohn and Vick, 2013, pp. 50–52, and Stacie L. Pettyjohn and Jennifer Kavanagh, *Access Granted: Political Challenges to the U.S. Overseas Military Presence, 1945–2014,* Santa Monica, Calif.: RAND Corporation, RR-1339-AF, forthcoming.

[77] DoD, 2012.

[78] Manyin et al., 2012, p. 3.

[79] Javier C. Hernandez and Floyd Whaley, "Philippine Supreme Court Approves Return of U.S. Troops," *New York Times,* January 12, 2016.

failed to come to specific terms for expanding USAF and USN presence, although they affirmed their intention to do so in a June 2014 force posture agreement.[80]

Limited progress is understandable—the pivot is a relatively recent development; major strategic changes are not implemented overnight; and small steps today can lay the foundation for bigger changes down the road. Yet with defense budget reductions and ongoing crises in the Middle East and Eastern Europe that require U.S. military forces, the Obama administration's rebalance appears to be more rhetoric than reality. The rebalance, therefore, is a cautionary tale that highlights the difficulty of making major changes to posture absent a catalytic event.[81]

Conclusions

Strategy clearly influences the shape and size of the U.S. global defense posture, but it appears to have less of an impact than many might assume. There is often a gap between the stated or desired strategy and what international and domestic audiences will actually support. Trying to make substantial changes to posture in a period of strategic ambiguity and ambivalence is likely to fail. Historically, major expansion of the U.S. overseas military presence has only occurred in periods of high threat—when new challenges are unambiguous and bring the United States and overseas partners together. During periods of extremely high threat, there is an overriding sense of urgency, which can help USAF overcome political and bureaucratic obstacles to establishing new bases. Therefore, a crisis or period in which domestic and international audiences perceive a significant threat often eases and allows strategic and deliberate modifications to U.S. posture. Similarly, major reductions in overseas presence have occurred rarely and only at the end of hot or cold wars. Existing bases have proven resistant to change, which will be discussed further in Chapter Five. Historically, different military strategies have resulted in frequent—although modest—changes to USAF's posture. In contrast, a different grand strategy could result in a dramatically different—and in all likelihood, much smaller—U.S. military presence overseas.

Strategy is an attempt to impose order and rationality in response to chaotic, complex, and difficult security challenges. Although significant domestic and international constraints prevent an idealized linking of means and ends, strategy, if nothing else, does help frame and articulate policy choices. In the next chapter, we consider what is essentially the polar opposite of strategy—the impact of contingent events on U.S. global posture.

---

[80] Cameron Stewart, "Go-Slow Signaled on Army Build Up," *The Australian*, November 14, 2012; Bruce Vaughn and Thomas Lum, "Australia: Background and U.S. Relations," Washington, D.C.: Congressional Research Service, December 14, 2015, pp. 1–2.

[81] The main point is to highlight how difficult it is to intentionally change posture to align it with the declared strategy when there is not a crisis or a clear threat. This is not intended to be a judgment about the appropriateness of the strategy of rebalancing. Given the challenges that have emerged in other regions since the rebalance was announced, the fact that the Obama administration has not been able to more significantly swing forces from Europe to the Asia may actually be a good thing.

4. Contingent Event Analysis

Introduction

To complement the previous discussion of deliberate planning and strategy development as driving mechanisms for stability or change in posture, this chapter considers the role of contingent, often unexpected events in posture development. Contingent events are arguably the most powerful driving mechanisms of posture because they can trigger demands for changes in strategy, presence, and basing and, as will be discussed in Chapters Five and Six, also initiate path dependent and path averting processes. Although contingency is a common and powerful phenomenon in human affairs (e.g., in scientific discoveries, financial market behavior, and international relations) the human cognitive system so desires order that chance is routinely downplayed or ignored and our individual and collective ability to predict events is vastly overstated. This can be seen whenever there is a major unexpected event in domestic politics, the economy, sports, or international affairs. After surprises, it is common to hear experts and laymen alike engage in logical acrobatics along the following lines: (1) The event was inevitable; (2) they had predicted it all along; or (3) their own (now demonstrably wrong) prediction or predictive system was accurate except for some variable not accounted for. Experts of all stripes are particularly adept at making such arguments after the fact.[82]

Although DoD planning processes consider a wide range of potential demands, these processes have significant limitations, particularly regarding potential events over a 30-year period. In particular, the planning processes are powerfully constrained by two types of bias. The first regards the likelihood of particular events. Assessing the probability of events is difficult to do in the short term; over a multidecade period, such judgments have little predictive value. Thus, events that are likely or at least plausible over a 30-year planning period are often ignored in institutional processes based on expert judgment. The second (and related) bias regards the importance of future events. Today's planners can reasonably assume that some classes of threats

---

[82]The most comprehensive assessment of the predictive performance of policy and political experts is Phillip Tetlock, *Expert Political Judgment: How Good Is It? How Can We Know?* Princeton, N.J.: Princeton University Press, 2005. For a broader treatment of bias and irrationality in human cognition, see Daniel Kahneman, *Thinking Fast and Slow*, New York: Farrar, Straus, and Giroux, 2011. On p. 215, Kahneman uses the example of Wall Street fund managers to illustrate our collective embrace of the false notion that experts can consistently predict future events: "for a large majority of fund managers, the selection of stocks is more like rolling dice than playing poker. Typically at least two out of every three mutual funds underperform the overall market in any given year." It should be noted that predicting market behavior is distinct from some types of gambling in closed systems with known odds. For more on the latter, see Nate Silver, *The Signal and the Noise: Why So Many Predictions Fail—But Some Don't*, New York: Penguin Press, 2012. Other related works on bias in human cognition relevant for military planners are Carol Tavris and Elliot Aronson, *Mistakes Were Made (But Not by Me): Why We Justify Foolish Beliefs, Bad Decisions and Hurtful Acts*, New York: Harcourt, 2007, and Daniel Gardner, *Future Babble: Why Pundits Are Hedgehogs and Foxes Know Best*, New York: Plume, 2012.

or demands considered most important to the United States in 2015 (e.g., deterrence of nuclear attack) will be among the highest priorities for the nation even several decades out.[83] In contrast, experts have almost no ability to predict where, when, and how the United States will decide to use military force. For example, no planner in 1971 envisioned U.S. military operations against the Taliban 30 years hence.

DoD has addressed some of these planning challenges through the use of planning scenarios and war gaming. Although most of these efforts are directed at specific challenges, Office of the Secretary of Defense Net Assessment and other organizations periodically cast the net much wider and farther out. The use of "wildcards" (low-probability, high-impact events) in scenario analysis and war gaming is helpful in recognizing critical junctures that overturn the status quo and lead to a new, very different steady state. In the same spirit, the concept of "Black Swan" events has been developed to capture high-impact, difficult-to-predict events.[84] Wildcard analysis may actually come closest to capturing the potential impacts on USAF posture from high-impact events, such as World War II or others discussed in this report. Most such planning activities, however, make judgments about the relative probability and importance of wildcards and dive deeply into one or a few that are judged most interesting or important.

In this analysis, we chose to go for breadth rather than depth for two reasons. First, as noted above, current DoD internal and DoD-funded scenario analysis tends toward deep consideration of a few possibilities. We judged that our research could be most helpful by pushing in the other direction, considering the impacts of a large number of events, including wildcards. Second, we judged that a broader assessment of scenarios would offer a test of the robustness of current posture planning across a wide range of big and small demands. Ideally, USAF and DoD would push planning in both directions: greatly expand the number of scenarios considered and deeply consider the implications of high-impact wildcards.

Therefore, as a modest step toward a more comprehensive consideration of potential demands in long-term posture planning, we developed a simple method to insert randomly selected operational vignettes into the analysis of posture demands. This method could be implemented by any planning staff, although ideally it would be supplemented by more-advanced and resource-intensive methods, such as massive scenario generation and Robust Decision Making.[85] This chapter describes our process and the results of the analysis.

---

[83] But even this seemingly straightforward assumption would be overturned if a defensive technology were developed that made nuclear weapons obsolete. This development is unlikely, but a long-range planner cannot rule out such possibilities.

[84] See Nassim Nicholas Taleb, *The Black Swan: The Impact of the Highly Improbable,* New York: Random House, 2010.

[85] These methods are described in Paul K. Davis, Steven C. Bankes, and Michael Egner, *Enhancing Strategic Planning with Massive Scenario Generation,* Santa Monica, Calif.: RAND Corporation, TR-392, 2007; Paul K. Davis, Russell D. Shaver, and Justin Beck, *Portfolio-Analysis Methods for Assessing Capability Options,* Santa Monica, Calif.: RAND Corporation, MG-662-OSD, 2008; and Robert J. Lempert, Steven W. Popper, and Steven C. Bankes, *Shaping the Next One Hundred Years: New Methods for Quantitative, Long-Term Policy Analysis,* Santa Monica, Calif.: RAND Corporation, MR-1626-RPC, 2003.

Operational Vignette Development

This analysis began by identifying 11 classes of conflict or challenge that could present demands on USAF forces and posture out to 2044. These classes are an update of those identified in a related 2009 RAND study that drew from a large number of scenario-generating documents and related activities.[86] One could certainly highlight particular challenges by adding classes (e.g., attacks on space assets); in our judgment, these 11 broad classes produce sufficient variation for analysis:

- disaster or epidemic
- civil conflict
- state failure
- government or regime change
- transnational conflict
- terrorism, piracy, and crime
- major cyberattack
- slow-intensity conflict (SLIC)
- limited conflict
- large-scale conflict
- nuclear use.

These conflict types and challenges are combined with 16 regions of the world to create a matrix.[87] With 11 columns and 16 rows, the matrix yields 176 potential operational demands, as shown notionally in Table 4.1 (the actual matrix is too large to display on an 8 by 11–inch page). We make no judgments about the relative importance or probability of these events; we have no illusions that we are any less biased than DoD planning staffs in making judgments about events three decades out, and low-probability, high-impact contingencies that current consensus deems as serious threats to U.S. interests are studied extensively and included in the formal operational planning process.

Large-scale conflict in North/Central Europe is an example of one operational demand. In 30 cases, the class of conflict–region intersection generated no sensible demand; for example, "civil conflict in Antarctica" is not applicable because there are no nations there within which civil strife could occur. The removal of nonsensical cases reduces the pool of potential demands to 146. In each of these 146 cases, our research team created an operational vignette drawn from historical experience, previous scenario analysis, current events, and expertise resident on our study team. We then identified missions (e.g., "airlift joint forces") that represent some of USAF

---

[86] For a related treatment of future challenges, see Frank Camm, Lauren Caston, Alexander C. Hou, Forrest E. Morgan, and Alan J. Vick, *Managing Risk in USAF Force Planning,* Santa Monica, Calif.: RAND Corporation, MG-827-AF, 2009.

[87] The regions are North/Central Europe, Southern Europe/East Mediterranean, Levant, Persian Gulf, North Africa, Sub-Saharan Africa, South Asia, Northeast Asia, East Asia (central), Southeast Asia, Oceania, Canada/United States/Mexico, Central America/Caribbean, South America, Arctic, and Antarctica.

operational demands that might occur if the vignette came to pass. These missions are not meant to be comprehensive or exclusive but are simply common missions that would often be conducted during that type of contingency. These are drawn from USAF historical experience in similar contingencies. The operational demands are stated in terms of selected airpower missions. We realize that the selection of one or a few missions may seem arbitrary or overly narrow, but this is necessary to access posture demands. Ideally, we would conduct a campaign-level analysis or war game for each vignette, but that level of detail would require vast resources that are not typically available to air staff planners or most studies.

Table 4.1. Vignette Generating Matrix

	Conflict type 1	Conflict type 2	...	Conflict type 11
Region 1	Vignette 1	Vignette 17	...	Vignette 161
Region 2	Vignette 2	Vignette 18	...	Vignette 162
...	...	...	...	...
Region 16	Vignette 16	Vignette 32	...	Vignette 176

Note that our operational vignettes are shorthand descriptors designed to capture the type and location of a challenge, but they are not fully formed narratives or scenarios. A scenario is typically understood among defense planners "as a postulated sequence of possible events with some degree of internal coherence, i.e., events associated with a 'story.'"[88] Since our goal is to greatly expand the number of potential events included in the analysis rather than dive deeply into any one event, the vignette (which could be considered a scenario sketch) is ideal.[89]

In the case of the aforementioned large-scale conflict in North/Central Europe, the operational vignette (#131) is Russia-NATO conflict in the Baltics. All 146 vignettes are shown in Tables 4.2 through 4.5. These regional distinctions resulted in some variation in the number of scenarios within each class, ranging from a low of seven for "nuclear use" to a high of 16 for "SLIC" and several other conflict categories.

A possible objection to this analytic approach needs to be addressed: The selections are not purely random because our research team generated the population from which the draw was conducted. We selected the 11 conflict categories, the 16 regions, and the vignettes that reflect

---

[88] Davis et al., 2007, p. 3. A somewhat different twist on the scenario concept is that of the alternative future or analytic case, such as one in which growth in demand for water greatly exceeds the baseline assumption in modeling of California water supply vulnerabilities. This is an example of the "scenario discovery" analytical technique used to identify "futures in which a policy performs unexpectedly well or poorly." See Andrew M. Parker, Sinduja V. Srinivasan, Robert J. Lempert, and Sandra H. Berry, "Evaluating Simulation-Derived Scenarios for Effective Decision Support," *Technological Forecasting and Social Change*, Vol. 91, 2015, p. 65.

[89] Any of these vignettes could be expanded into a military planning scenario.

the intersection of a conflict category and region. There is no way to entirely escape human bias in this process. The best that one can do is use a large number of conflict categories.

Table 4.2. Disaster or Epidemic, Civil Conflict, and State Failure Vignettes

Number	Vignette	Type	Location
1	Flu epidemic in Sweden	Disaster/Epidemic	North/Central Europe
2	Mt Vesuvius erupts	Disaster/Epidemic	Southern Europe/East Med
3	Aleppo earthquake	Disaster/Epidemic	Levant
4	MERS epidemic in Saudi Arabia	Disaster/Epidemic	Persian Gulf
5	Tunisian earthquake	Disaster/Epidemic	North Africa
6	Ebola in Ethiopia	Disaster/Epidemic	Sub-Saharan Africa
7	Bangladesh floods	Disaster/Epidemic	South Asia
8	Tsunami hits Japan	Disaster/Epidemic	Northeast Asia
9	Taiwan Typhoon	Disaster/Epidemic	East Asia (central)
10	Indonesian earthquake	Disaster/Epidemic	Southeast Asia
11	Cyclone hits Tonga	Disaster/Epidemic	Oceania
12	Tsunami hits Mexico	Disaster/Epidemic	Canada, US, Mexico
13	Hurricane hits Cuba	Disaster/Epidemic	Central America/Carb
14	Earthquake Chile	Disaster/Epidemic	South America
15	Russian oil well explodes	Disaster/Epidemic	Arctic
16	Major oil accident	Disaster/Epidemic	Antarctica
17	Ukraine	Civil conflict	North/Central Europe
18	Kosovo	Civil conflict	Southern Europe/East Med
19	Jordan (with Palestinian population)	Civil conflict	Levant
20	Shia uprising in Bahrain	Civil conflict	Persian Gulf
21	Mali	Civil conflict	North Africa
22	Nigeria	Civil conflict	Sub-Saharan Africa
23	Afghanistan	Civil conflict	South Asia
24	DPRK	Civil conflict	Northeast Asia
25	Myanmar	Civil conflict	Southeast Asia
26	Solomon Islands	Civil conflict	Oceania
27	Mexico	Civil conflict	Canada, US, Mexico
28	El Salvador	Civil conflict	Central America/Carb
29	Colombia	Civil conflict	South America
30	Ukraine	State failure	North/Central Europe
31	Kosovo	State failure	Southern Europe/East Med
32	Lebanon	State failure	Levant
33	Saudi Arabia	State failure	Persian Gulf
34	Libya	State failure	North Africa
35	Nigeria	State failure	Sub-Saharan Africa
36	Pakistan	State failure	South Asia
37	DPRK	State failure	Northeast Asia
38	China	State failure	East Asia (central)
39	Cambodia	State failure	Southeast Asia
40	Solomon Islands	State failure	Oceania
41	Mexico	State failure	Canada, US, Mexico
42	Haiti	State failure	Central America/Carb
43	Venezuela	State failure	South America

NOTE: MERS = Middle East Respiratory Syndrome

The conflict categories we chose, although broad, were inherently biased by our understanding of contemporary security challenges, as are the specific vignettes identified within these categories. A related exercise conducted in the 1990s would not have included "major cyberattack." Presumably, a similar analysis ten years hence will include categories and vignettes that we missed. For example, we have no cases in which kinetic attacks are made on space assets. These could occur under the "limited conflict" class, but future analysts might see the necessity for a separate "space war" class. Future studies using this technique may wish to be more aggressive and speculative in the types of conflict explicitly considered.

The large number of conflict classes and regions does generate many more vignettes than typically considered in planning processes, including many that defense planners would not identify as probable or important. Indeed, when our research team first saw the results of the random draw, we were disappointed that many of the vignettes were not "interesting," a validation of the need for and value of random vignette or scenario generation in long-term defense planning and analysis. This approach, while admittedly imperfect, greatly expands the scope of challenges considered and forces planners to consider possibilities that would be rejected due to bias and the inherent limits to human knowledge about future events.

Another possible criticism is that the draw was overly random because it did not weight probabilities or the importance of events. Thus, in the draws that we did, no wildcard events occurred; as a result, the posture implications of the vignettes were relatively modest. This is an interesting problem. Future studies might want to build a larger population of wildcard events and do a separate random draw from them, thus ensuring the consideration of at least one posture-shaking contingency.[90]

Most of these problems can be addressed in future studies by employing massive scenario generation and similar analytic techniques that allow the consideration of thousands of scenarios and the interaction of complex processes. Since our study sought to develop a method executable by air staff offices, we kept to spreadsheet-based methods, although wildcard draws and a larger vignette population could both be supported with a simple Excel spreadsheet.

---

[90] Our reviewers were split on these issues. One raised the problem of researcher bias in the selection of vignettes; the other was concerned about the lack of high-impact events in the analysis.

Table 4.3. Government/Regime Change, Transnational Conflict, and Terrorism/Piracy/Crime Vignettes

44	Germany	Govt/Regime change	North/Central Europe
45	Italy	Govt/Regime change	Southern Europe/East Med
46	Jordan	Govt/Regime change	Levant
47	Qatar	Govt/Regime change	Persian Gulf
48	Djibouti	Govt/Regime change	North Africa
49	South Africa	Govt/Regime change	Sub-Saharan Africa
50	Iran	Govt/Regime change	South Asia
51	Korea	Govt/Regime change	Northeast Asia
52	China	Govt/Regime change	East Asia (central)
53	Philippines	Govt/Regime change	Southeast Asia
54	Micronesia	Govt/Regime change	Oceania
55	Mexico	Govt/Regime change	Canada, US, Mexico
56	Cuba	Govt/Regime change	Central America/Carb
57	Colombia	Govt/Regime change	South America
58	IE in Albania	Transnational conflict	Southern Europe/East Med
59	IE in Syria & Iraq	Transnational conflict	Levant
60	AQAP moves into Oman	Transnational conflict	Persian Gulf
61	IE in Libya	Transnational conflict	North Africa
62	Boko Harem Nigeria	Transnational conflict	Sub-Saharan Africa
63	Taliban in Pakistan	Transnational conflict	South Asia
64	IS in Indonesia	Transnational conflict	Southeast Asia
65	US-Mexico border	Transnational conflict	Canada, US, Mexico
66	El Salvador drug cartels	Transnational conflict	Central America/Carb
67	Drug cartels Colombia	Transnational conflict	South America
68	Frequent terror attacks in France	Terrorism/piracy/crime	North/Central Europe
69	Soccer stadium bombing in Germany	Terrorism/piracy/crime	North/Central Europe
70	US Embassy in Jordan hit by NLOS missile	Terrorism/piracy/crime	Levant
71	AQAP attacks US embassy in Oman	Terrorism/piracy/crime	Persian Gulf
72	Airliner shot down Cairo airport	Terrorism/piracy/crime	North Africa
73	Nigerian offshore wells seized	Terrorism/piracy/crime	Sub-Saharan Africa
74	Mubai style attacks in Surat, India	Terrorism/piracy/crime	South Asia
75	Hikari no Wa sarin attack Tokyo subway	Terrorism/piracy/crime	Northeast Asia
76	Uighur Salafi truck bomb in Beijing	Terrorism/piracy/crime	East Asia (central)
77	Return of Malaccan Strait piracy	Terrorism/piracy/crime	Southeast Asia
78	Chinese fishermen killed in Micronesia	Terrorism/piracy/crime	Oceania
79	Terrorist attack Washington DC	Terrorism/piracy/crime	Canada, US, Mexico
80	Organized crime threats to El Savador gov	Terrorism/piracy/crime	Central America/Carb
81	Terrorists bomb Colombian parliament	Terrorism/piracy/crime	South America
82	Suspicious explosion at US oil well	Terrorism/piracy/crime	Arctic
83	Ecotourism ship taken hostage	Terrorism/piracy/crime	Antarctica

Table 4.4. Major Cyber Attack, SLIC, and Limited Conflict Vignettes

84	Major accident Berlin subway-Russian hack	Major cyber attack	North/Central Europe
85	London banks hacked for $billions-Russia	Major cyber attack	Southern Europe/East Med
86	Tel Aviv water system shutdown-unk	Major cyber attack	Levant
87	Qatar natural gas storage explosion-Iran hack	Major cyber attack	Persian Gulf
88	Algerian oil distribution system shutdown-IS	Major cyber attack	North Africa
89	Cape Town power system damaged-unk	Major cyber attack	Sub-Saharan Africa
90	Pakistan military comm taken down-India	Major cyber attack	South Asia
91	Hamaoka nuclear plant meltdown-DPRK	Major cyber attack	Northeast Asia
92	Taiwan flghter aircraft lost due to hack-PRC	Major cyber attack	East Asia (central)
93	Singapore midair collision due to ATC hack	Major cyber attack	Southeast Asia
94	New Zealand stock market hack-unk	Major cyber attack	Oceania
95	US telecomm severe disruption-Iran	Major cyber attack	Canada, US, Mexico
96	Panama Canal closed due to hack-unk	Major cyber attack	Central America/Carb
97	Brazilian banking system closed for weeks-unk	Major cyber attack	South America
98	American oil well explodes-Russian hack	Major cyber attack	Arctic
99	All power lost at McMurdo station-unk	Major cyber attack	Antarctica
100	Russian AF in Baltic	SLIC	North/Central Europe
101	Turkey-Cyprus-Israel oil reserves	SLIC	Southern Europe/East Med
102	Israel-Lebanon over oil	SLIC	Levant
103	Iran harasses PG shipping	SLIC	Persian Gulf
104	Morocco-Algeria border dispute	SLIC	North Africa
105	DRC-Rwanda border dispute	SLIC	Sub-Saharan Africa
106	China-India maritime	SLIC	South Asia
107	DPRK-ROK maritime	SLIC	Northeast Asia
108	East China Sea oil fields	SLIC	East Asia (central)
109	Chinese expansion SCS	SLIC	Southeast Asia
110	Chinese illegal fishing	SLIC	Oceania
111	US-Mexico over fisheries	SLIC	Canada, US, Mexico
112	Nicaragua-Honduras fisheries	SLIC	Central America/Carb
113	Venezuela-Colombia maritime	SLIC	South America
114	US-Russia over oil	SLIC	Arctic
115	China-Chile over oil	SLIC	Antarctica
116	Russian-Swedish navies	Limited conflict	North/Central Europe
117	Turkey-Russia naval/air	Limited conflict	Southern Europe/East Med
118	Jordan-Syria	Limited conflict	Levant
119	Saudi Arabia-Bahrain	Limited conflict	Persian Gulf
120	Israel-Egypt in Sinai	Limited conflict	Levant
121	Sudan invades S Sudan	Limited conflict	Sub-Saharan Africa
122	India-Pakistan	Limited conflict	South Asia
123	Japan-Russia Kurile Islands	Limited conflict	Northeast Asia
124	China lands forces Senkakus	Limited conflict	East Asia (central)
125	China-Philippines, Spratlys	Limited conflict	Southeast Asia
126	Chinese intervene in Micronesia	Limited conflict	Oceania
127	Nicaragua-El Salvador	Limited conflict	Central America/Carb
128	Venezuela-Colombia	Limited conflict	South America
129	Russia-Canada over oil	Limited conflict	Arctic
130	Chile-Peru over fisheries	Limited conflict	Antarctica

NOTE: SCS = South China Sea

Table 4.5. Large-Scale Conflict and Nuclear Use Vignettes

131	Russia-NATO Baltics	Large-scale conflict	North/Central Europe
132	Turkey/NATO vs Russia	Large-scale conflict	Southern Europe/East Med
133	Israel-Syria	Large-scale conflict	Levant
134	Iran-Saudi Arabia	Large-scale conflict	Persian Gulf
135	Libya-Egypt	Large-scale conflict	North Africa
136	India-Pakistan	Large-scale conflict	South Asia
137	DPRK-ROK	Large-scale conflict	Northeast Asia
138	China-Taiwan	Large-scale conflict	East Asia (central)
139	China-Vietnam	Large-scale conflict	Southeast Asia
140	Russian EMP over Kiev	Nuclear use	North/Central Europe
141	Iran threatens Israel	Nuclear use	Levant
142	Dirty bomb Baghdad	Nuclear use	Persian Gulf
143	Pakistan strikes India	Nuclear use	South Asia
144	DPRK threatens ROK	Nuclear use	Northeast Asia
145	PRC threatens US	Nuclear use	East Asia (central)
146	Dirty bomb in LA	Nuclear use	Canada, US, Mexico

The next step in our analysis used the Excel random number function to generate 30 random numbers, one for each of the 30 years in our planning period. This generates the vignette list found in Table 4.6. As might be expected when making 30 random selections from a population of 146 (sample size of roughly 20 percent of population), some vignettes appeared multiple times. For example, the "MERS epidemic in Saudi Arabia" occurred three times in the 30-year period, and the Libya-Egypt conflict occurred twice. The reoccurrence of a problem within a three-decade period is not entirely unrealistic but does somewhat reduce the breadth of cases for analysis. For future studies, it may be worth using countries instead of regions, which would generate over ten times as many cases. This approach would reduce vignette duplication and also might partially address the population bias problem discussed earlier.[91]

---

[91] If United Nations member states were used instead of our 16 regions to generate scenarios, the total number of potential scenarios would increase from 176 to 2,123 (193 member states multiplied by 11 conflict or challenge classes). A more ambitious approach could use a computer model to directly generate large numbers of vignettes or scenarios using such characteristics as location, type of conflict, identity of adversaries, date, level of armaments, and competence of combatants. This is in the spirit of the massive scenario generation envisioned in Davis et al., 2007.

Table 4.6. Vignettes Selected Using Random Number Generator

Time period	Vignette	Conflict Type	Random Number
2015-2024	Cuba	Govt/Regime change	56
2015-2024	Shia uprising in Bahrain	Civil conflict	20
2015-2024	Colombia	Civil conflict	29
2015-2024	Germany	Govt/Regime change	44
2015-2024	Libya-Egypt	Large-scale conflict	135
2015-2024	Nigerian offshore wells seized	Terrorism/piracy/crime	73
2015-2024	Drug cartels Colombia	Transnational conflict	67
2015-2024	China lands forces Senkakus	Limited conflict	124
2015-2024	US-Russia over oil	SLIC	114
2015-2024	Iran-Saudi Arabia	Large-scale conflict	134
2025-2034	Turkey/NATO vs Russia	Large-scale conflict	132
2025-2034	Russian oil well explodes	Disaster/Epidemic	15
2025-2034	Hikari no Wa sarin attack Tokyo subway	Terrorism/piracy/crime	75
2025-2034	South Africa	Govt/Regime change	49
2025-2034	Libya-Egypt	Large-scale conflict	135
2025-2034	MERS epidemic in Saudi Arabia	Disaster/Epidemic	4
2025-2034	Terrorists bomb Colombian parliament	Terrorism/piracy/crime	81
2025-2034	Kosovo	Civil conflict	18
2025-2034	Israel-Egypt in Sinai	Limited conflict	120
2025-2034	Jordan-Syria	Limited conflict	118
2035-2044	Pakistan military comm taken down-India	Major cyber attack	90
2035-2044	Saudi Arabia	State failure	33
2035-2044	All power lost at McMurdo station-unk	Major cyber attack	99
2035-2044	Drug cartels Colombia	Transnational conflict	67
2035-2044	Brazilian banking system closed for weeks	Major cyber attack	97
2035-2044	MERS epidemic in Saudi Arabia	Disaster/Epidemic	4
2035-2044	MERS epidemic in Saudi Arabia	Disaster/Epidemic	4
2035-2044	US-Mexico over fisheries	SLIC	111
2035-2044	Tsunami hits Mexico	Disaster/Epidemic	12
2035-2044	Korea	Govt/Regime change	51

Six of the vignettes listed in Table 4.6 had no obvious USAF mission and were dropped from the analysis.[92] As noted above, there also were duplicate scenarios: The Libya-Egypt vignette

---

[92] There were four government or regime change vignettes that did not generate direct operational demands: Cuba, Germany, South Africa, and Korea. On the other hand, two of them (Korea and Germany) postulated radical changes in government that forced the United States to remove all forces from their respective territories. These two vignettes would be highly disruptive to USAF posture and would likely have lasting negative impacts, particularly the loss of basing in Germany. That said, none of these vignettes generate immediate operational demands. Two other vignettes (Pakistani and Brazilian cyberattacks) also failed to generate operational demands.

appeared twice, and the MERS vignette appeared three times. After subtracting the "not applicable" vignettes and the duplicates, 21 unique events were left for analysis.

We next identified representative missions for each of the 21 remaining vignettes, as shown in Table 4.7. For example, the third vignette in the list, Libya-Egypt war, envisions Egyptian Air Force (EAF) fighter aircraft conducting strikes against targets in Libya. The representative USAF mission we selected was air refueling of the EAF fighters using KC-135 aircraft. We selected Benghazi as the rough location for the tanker orbits to provide a geographic center for the analysis. We used prior RAND analyses, USAF documents, and the operational expertise of a USAF RAND fellow to determine the preferred maximum operating distances and minimum operating surface requirements for each mission. In the case of the tankers, the preferred operating distance was 700 nm to maximize fuel offload, and the preferred airfield would have runway dimensions of 10,500 by 147 ft and a Pavement Classification Number (PCN) of 60. A few vignettes (e.g., United States–Russia over oil) list multiple missions and aircraft. For the illustrative analysis, we chose just one of these missions and one aircraft type. The aircraft type selected for analysis, as well as aircraft range and runway requirements, are shown in red for the six vignettes that involve multiple missions or platforms.

Table 4.7. From Vignettes to Missions and Posture Requirements

Vignette	Key missions	Aircraft	Range	Geographic Center	Runway (ft)
Shia uprising in Bahrain	ISR	MQ-9	500nm	Manama	5000
Colombia	ISR/CAS	AC-130	200nm	Bogota	3000 X 60
Libya-Egypt	Air refueling for EAF	KC-135	700nm	Benghazi	10,500 X 147
Nigerian offshore wells seized	ISR	MQ-9	500nm	Tip of Delta	5000
Drug cartels Colombia	ISR	E-3	200nm	Cali	9000 X 147
China lands forces Senkakus	Air superiority	F-22	1500nm	Senkakus	7500
US-Russia over oil	ISR	P-8, RQ-4	1700nm	Chukchi Sea	8000
Iran-Saudi Arabia	Destroy TBM TELs	E-8, F-22	300-1500nm	> 300nm from Iranian shoreline; <1500nm from launch arc in Iran*	7500
Turkey/NATO vs Russia	Air superiority, maritime strike	E-3, P-8, F-22, F-35	1500nm	~Center Black Sea	7500
Russian oil well explodes	Airlift oil spill materials	C-17	4400nm	Nome	3000 X 90
Sarin attack Tokyo subway	Airlift medical support	C-17	4400nm	Tokyo	3000 X 90
Libya-Egypt	Air refueling for EAF	KC-135	700nm	Benghazi	7000 X 147
MERS epidemic in Saudi Arabia	Airlift	C-17	4400nm	Riyadh	3000 X 90
Bombing of Colombian parliament	Airlift	C-17	4400nm	Bogota	3000 X 90
Kosovo	ISR	MQ-9	500nm	Pristina	5000
Israel-Egypt in Sinai	ISR	RQ-4	1700nm	El Arish	8000
Jordan-Syria	Strike	MQ-9	500nm	El Taebah, Syria	5000
Saudi Arabia	NEO, ISR, airlift joint forces	C-17, MQ-9	2000nm, 500nm	Riyadh	5000
Power lost at McMurdo station	Airift	C-17	2000nm	McMurdo Station	3000 X 90
Drug cartels Colombia	ISR	E-3	200nm	Cali	9000 X 147
MERS epidemic in Saudi Arabia	Airlift	C-17	4400nm	Riyadh	3000 X 90
MERS epidemic in Saudi Arabia	Airlift	C-17	4400nm	Riyadh	3000 X 90
US-Mexico over fisheries	ISR	MQ-9, RQ-4	500nm/1700nm	200nm S of Texas/Louisana border	5000
Tsunami hits Mexico	Airlift, ISR	C-17, MQ-9, RQ-4	2000nm, 500nm	Acapulco	5000

Finally, we used the Automated Air Facility Information File (AAFIF) and geographic information systems software (ArcGIS) to identify airfields that met mission requirements.[93] We used this software to create the appropriate range buffer around each geographic center. For each scenario, the buffer was intersected with the airfields listed in the AAFIF that matched runway and PCN requirements for aircraft. The result is shown in Figure 4.1. Only 13 airfields in eight countries met these demanding conditions.

Figure 4.1. Airfields That Meet Requirements for Egypt-Libya Conflict Vignette

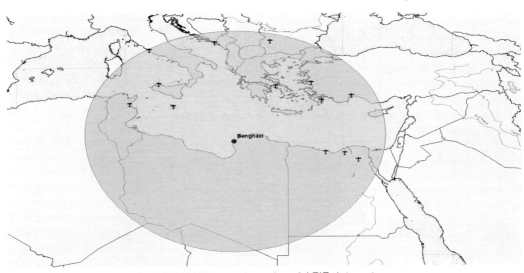

SOURCE: RAND map based on AAFIF data set.

If, however, we allow the tankers to operate from farther distances and/or with less-conservative runway requirements, the number of potential airfields and partner nations increases greatly. For example, the Air Force instruction for the KC-135 gives 7,000 ft as the minimum runway length.[94] Although aircraft weight, air temperature, humidity, winds, and airfield altitude will often require a longer operating surface, there are times when a 7,000 ft runway would be adequate. If the analysis is done with that more relaxed runway length assumption, the AAFIF identifies 41 airfields in 13 countries (see Figure 4.2). Similarly, allowing less-than-optimal fuel offloads would also increase the number of potential airfields. Most real-world operations will fall somewhere between these two extremes.

---

[93] National Geospatial Intelligence Agency Defense Mapping Agency, Automated Air Facility Information File, data set. Not available to general public.

[94] Air Force Instruction 11-2 KC-135, *Flying Operations, C/KC-135 Aircraft Configuration,* Vol. 3, March 6, 2015, p. 80.

Figure 4.2. Airfields That Meet Requirements for Egypt-Libya Conflict Vignette Under Relaxed Airfield Assumptions

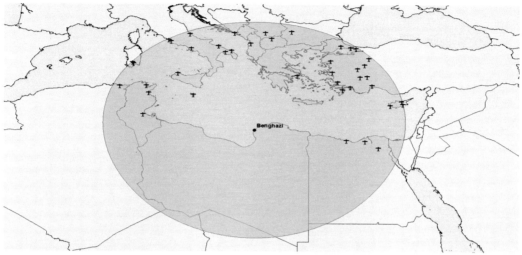

SOURCE: RAND map based on AAFIF data set.

The contingent event analysis conducted similar calculations for all vignettes in Table 4.6. Once all 21 vignettes were assessed, we compared participation by country (i.e., the number of vignettes in which a given country's airfields met operational criteria). We then calculated average participation across countries (country mean) in each region (the average number of scenarios countries participated in). For each region, we also calculated the highest number of vignettes that any one country participated in (country max) and the total number of vignettes covered by that region (region max) as seen in Figure 4.3.

For this analysis, airfields in Europe and the Middle East were most versatile. Europe has the highest single country score, with Greece participating in nine vignettes, and the highest maximum score of ten—the number of vignettes in which at least one European country could offer operationally relevant airfields. The Middle East had the highest average (mean) score, with each country participating in approximately five vignettes. The Western Hemisphere scored next best, with Africa and Asia the lowest performers. Asia scored lowest for two reasons. First, the random draw selected only one scenario in Northeast Asia (sarin in Tokyo), only one in East Asia central (China/Senkakus), none in Southeast Asia, and only one in South Asia (India-Pakistan). A different draw could easily have included more Asia vignettes. Second, and more important, the Asian airfields are generally too remote to be useful for operations in the Middle East, Europe, and the Western Hemisphere. The Antarctic McMurdo Station vignette was the exception; airlift flights generally launch from New Zealand.

It is interesting that this result regarding the utility of airfields as forward operating locations is consistent with an earlier RAND analysis measuring the versatility of 35 airfields as en route locations for airlift across 28 diverse scenarios. Airfields in Europe and the Middle East scored

43

highest in that analysis as well, with the Western Hemisphere locations scoring lowest of any region.[95]

Figure 4.3. Vignette Participation by Region

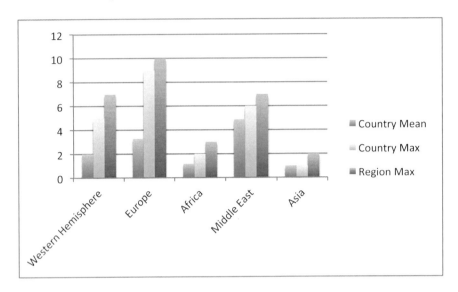

The better performance of airfields in Europe and the Middle East could be seen as an artifact of the way the analysis divided the world into 16 regions and of the particular vignettes in this random draw. Geography may also play a role. In the Mediterranean littoral, Europe, Africa, and the Middle East all intersect in a manner that allows airfields in any one region to support air operations in the other two. Africa scores low because there are relatively few quality airfields; if airfields were improved in northern and eastern Africa, its scores would be much higher.

Finally, for comparison, Figure 4.4 displays the total number of participating countries. Europe easily dominated on this metric, with 26 countries contributing to at least one vignette. Note that there is considerable overlap in coverage across vignettes. Thus, although 26 countries were included, they only cover a total of 10 vignettes (as noted above).

---

[95] Pettyjohn and Vick, 2013, pp. 30–34.

Figure 4.4. Number of Vignette Participating Countries by Region

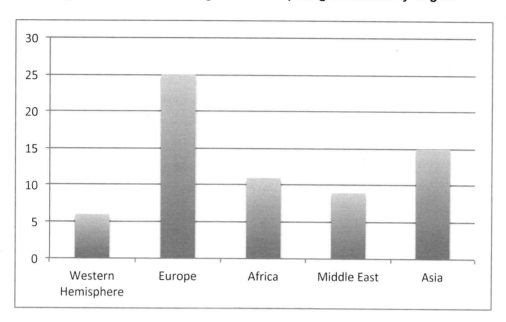

In addition to the 21 vignettes that generated immediate operational demands on USAF force structure and posture, there were two vignettes that—although creating no immediate military missions had major long-term implications for USAF posture. In the Germany and Korean government and regime change vignettes, the new governments asked U.S. forces to leave their respective countries. The German and Korean vignettes are admittedly extreme, but that is the point of the random draw: to force consideration of events that are deemed highly improbable.

Loss of access to airfields (and military facilities more broadly) in Germany would be highly disruptive to U.S. strategy and military operations, mainly because of the quality, number, and size of facilities rather than geography per se. For example, in our analysis, German airfields contributed to only two vignettes. Germany was outscored by ten other European countries, including Italy, with seven vignettes. Consistent with this, in the 2013 RAND report cited earlier, airfields in Turkey, Greece, Italy, and Spain all scored as high as German airfields for airlift versatility.[96]

In the Korean case, the posture implications are more limited. This is because the location of the Korean airfields in Northeast Asia limits their utility for operations off the Korean Peninsula. Perhaps more important, although less certain, is the generally held view that Korean airfields would not be available for use by U.S. forces except in the defense of South Korea (or for benign missions, such as humanitarian relief).[97]

---

[96] Pettyjohn and Vick, 2013, p. 33.

[97] For more on the limited versatility of Korean airfields, see Pettyjohn and Vick, 2013, p. 33.

Implications for USAF Posture

The analysis in this chapter sought to complement the more deliberative strategy-based posture assessment presented in Chapter Three by exploring the implications of contingent events for USAF posture. As discussed earlier, incorporating randomly chosen scenarios forces planners to consider potential long-term demands that current biases rule out as unlikely, unimportant, or both. Doing so has great value as a supplement to other analytical and planning techniques as a test of the robustness of USAF posture and plans. In this particular analysis, USAF global posture proved robust. In every vignette, existing USAF facilities could conduct the operation, or the vignette took place entirely within the host country (such as in the Nigerian and Colombian cases, where the United States would contribute forces at the invitation of those governments). In most vignettes, potential operating locations existed in multiple nations (many of which are current U.S. security partners).

Contingent event analysis does have limitations, however, including potential bias in the definition of conflict types, regions, and selection of vignettes for the random draw. Surveys, Delphi-type exercises, crowd sourcing, gaming, and computer generation of vignettes might all be used to expand the scenario space in future studies, reducing the impact of expert bias or analyst lack of imagination.

In this chapter, we have shown how contingent events can generate immediate demands on USAF global posture or reduce future access. These events also can set in train a process in which initial conditions rapidly establish a path that is difficult to depart from. This phenomenon of "path dependence" has great significance for USAF posture and is the topic of the next chapter.

5. Path Dependence in USAF Posture

This chapter describes and assesses the influence of path dependence, the third driver of posture considered in this report. We begin by defining path dependent processes. The second section then discusses how path dependence might affect posture. The third section explores whether there is evidence to suggest that U.S. posture is path dependent at four different levels. The fourth offers some conclusions.

Background

Testifying before the Senate Armed Services Committee in 2004, Secretary of Defense Donald Rumsfeld noted that U.S. forces overseas "are still situated in a large part as if little has changed for the last 50 years—as if, for example, Germany is still bracing for a Soviet tank invasion across its northern plain." Similarly, in South Korea, it appeared as if U.S. "troops were virtually frozen in place from where they were when the Korean War ended in 1953."[98] The tendency for a U.S. overseas military presence to persist is not a new phenomenon. It dates back to 1815, when the first American forces were indefinitely deployed overseas—the USN Mediterranean squadron—to protect U.S. commerce from the Barbary pirates. Within several months, the Mediterranean squadron defeated the pirates, but because the general threat of privateering lingered until the French conquered Algeria in 1830, USN forces remained in the region. But by the 1830s, "the policy of having a Mediterranean squadron had become so well fixed that it was able to go on of its own momentum during the nineteenth century."[99] Therefore, despite the absence of a threat or immediate need, the squadron, which was rebranded the European squadron in 1865, endured until 1905.

These examples raise an important question: Why does the U.S. military presence overseas often seem resistant to change? An overseas base is usually established for a specific purpose—to deal with a new threat or to carry out a particular operation. One would expect that when the threat disappeared or the operation ended, the base would be closed and U.S. forces would return home. Yet, as Rumsfeld pointed out, it is quite common for the United States to maintain a base long after its original purpose is gone. What explains this inertia?

Some have argued that the persistence of U.S. bases is evidence that the United States seeks to dominate the world; others suggest that its network of overseas bases is a form of informal

---

[98] Donald Rumsfeld, "Prepared Testimony of the U.S. Secretary of Defense Donald H. Rumsfeld Before the Senate Armed Services Committee: Global Posture," September 23, 2004, p. 4.

[99] Robert G. Albion, "Distant Stations," *U.S. Naval Institute Proceedings,* Vol. 80, March 1954, pp. 265–273; see also Harold Hance Sprout and Margaret Tuttle Sprout, *The Rise of American Naval Power, 1776–1918,* Annapolis, Md.: Naval Institute Press, 1990 (originally published by Princeton University Press, 1939, 1966), pp. 117–119.

empire.[100] Although U.S. acquisition of overseas territories in the late 19th century was a form of empire building, nearly all U.S. overseas bases since 1945 have been obtained with the agreement of the host nation. As a result, U.S. forces leave when they are asked, cannot dictate the terms of basing agreements, and cannot force a country to permit it to use its bases for a particular mission.[101]

Something much more mundane may help explain why a U.S. military presence often persists well beyond its original purpose—path dependence—the idea that once a choice is made (often for very contingent reasons) that it quickly becomes locked in and difficult to change. The remainder of this chapter explores that possibility.

Path Dependent Processes

Economists first developed the concept of path dependence to explain technological adoption and, in particular, suboptimal or inefficient outcomes.[102] Since that time, the idea has been adapted and applied to the fields of sociology and political science to explain institutional persistence. At the most basic level, path dependence means that "history matters" or that the "past affects the future."[103] Although there are differences of opinion about exactly what path dependence means, most scholars agree that it has four defining characteristics (see Figure 5.1):[104]

1. openness
2. a critical juncture characterized by contingency
3. constraint
4. closure.

---

[100] David Vine, *Island of Shame: The Secret History of the U.S. Military Base on Diego Garcia*, Princeton, N.J.: Princeton University Press, 2012; Catherine Lutz, ed., *The Bases of Empire: The Global Struggle Against U.S. Military Posts*, New York: New York University Press, 2009; Mark L. Gillem, *America Town: Building the Outposts of Empire*, Minneapolis: University of Minnesota Press, 2007; and Joseph Gerson and Bruce Birchard, eds., *The Sun Never Sets: Confronting the Network of Foreign U.S. Military Bases*, Boston: South End Press, 1991.

[101] Alexander Cooley and Hendrik Spruyt, *Contracting States: Sovereign Transfers in International Relations*, Princeton, N.J.: Princeton University Press, 2009; Alexander Cooley and Daniel Nexon, "The Empire Will Compensate You': The Structural Dynamics of the U.S. Overseas Basing Network," *Perspectives on Politics*, Vol. 11, No. 4, December 2013, pp. 1039–1042.

[102] See Brian Arthur, *Increasing Returns and Path Dependence in the Economy*, Ann Arbor: University of Michigan Press, 1994.

[103] James Mahoney and Daniel Schensul, "Historical Context and Path Dependence," in Robert E. Goodwin and Charles Tilly, eds., *The Oxford Handbook of Contextual Political Analysis*, Oxford, UK: Oxford University Press, March 2006, p. 3; Paul Pierson, "Increasing Returns, Path Dependence, and the Study of Politics," *American Political Science Review*, Vol. 94, No. 2, June 2000b, p. 252.

[104] Andrew Bennett and Colin Elman, "Complex Causal Relations and Case Study Methods: The Example of Path Dependence," *Political Analysis*, Vol. 14, No. 3, Summer 2006, pp. 252–253.

Figure 5.1. Path Dependence

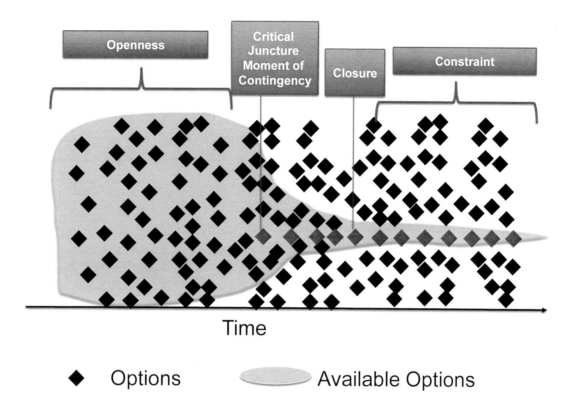

First, openness is the notion that at a point in time more than one path could have been chosen.[105] Path dependent processes are initially characterized by openness; therefore, the outcome is not predetermined or inevitable. Instead, in this early phase, a range of options is available, and the ultimate outcome depends on which option is chosen from the many possibilities.

Second, during a critical juncture—a key turning point characterized by uncertainty—one option is selected from the many possibilities for contingent reasons, and this choice sets a course that is difficult to change. A critical juncture is "a period of significant change" that "produce[s] distinct legacies."[106] An important feature of critical junctures is the fact that a key decision is made for random or highly contingent reasons or that it cannot be accounted for by prevailing theories.[107] The logic of critical junctures suggests that timing—when an event occurs in a sequence—can determine how it affects outcomes. In particular, there is an inverse

---

[105] Bennett and Elman, 2006, p. 252; Paul Pierson, "Not Just What, But When: Timing and Sequence in Political Processes," *Studies in American Political Development*, Vol. 14, Spring 2000a, p. 75.

[106] David A. Collier and Ruth Collier, *Shaping the Political Arena: Critical Junctures, the Labor Movement, and Regime Dynamics in Latin America,* Princeton, N.J.: Princeton University Press, 1991, p. 29.

[107] James Mahoney, "Path Dependence in Historical Sociology," *Theory and Society*, Vol. 29, No. 4, August 2000, p. 513; Mahoney and Schensul, 2006, p. 4; Bennett and Elman, 2006, p. 254.

relationship between when an event occurs and its influence, so events that take place earlier have outsized consequences.[108]

Third, there are constraints that "lock in" the decision made at the critical juncture, making it increasingly difficult to shift paths without paying very high costs.[109] The costs of reversing the earlier decision increase because of self-reinforcing positive feedback that encourages stability. It is important to note that the mechanisms of reproduction differ from the factors that led to the initial decision.[110] There are four major types of self-reinforcing processes: increasing returns, functional, power, and legitimacy beliefs. Each of these self-reinforcing processes has a different feedback mechanism that leads to the reproduction of the decision that was made at a critical juncture.[111] These are summarized in Table 5.1.

Table 5.1. Path Dependent Feedback Mechanisms

Self-Reinforcing Process	Mechanism of Reproduction	Mechanism of Change
Increasing returns	Decision is reproduced because of rational cost-benefit analysis	Costs of changing or benefits of current path decline
Functional	Decision is reproduced because it is increasingly useful to system as a whole	Declining usefulness
Legitimacy beliefs	Decision is reproduced because it is seen as the correct or appropriate course of action	Changing perceptions of what is appropriate
Power	Decision is reproduced because it is supported by elite actors	Changes in balance of power that weaken elites or empower subordinates

The concept of increasing returns comes from the economic literature and the idea that each movement in a particular direction increases the costs of changing paths and/or the benefits of the current path; therefore, rational decisionmakers elect to stay the course. This is modeled

---

[108] Mahoney, 2000, p. 510; Pierson, 2000b, p. 263; Taylor C. Boas, "Conceptualizing Continuity and Change: The Composite-Standard Model of Path Dependence," *Journal of Theoretical Politics*, Vol. 19, No. 1, 2007 p. 37; Paul Pierson, *Politics in Time: History Institutions, and Social Analysis*, Princeton, N.J.: Princeton University Press, 2004, p. 67.

[109] Pierson, 2000b, p. 252.

[110] Mahoney, 2000, p. 515.

[111] Mahoney, 2000, pp. 515–526; Kathleen Thelen, "Historical Institutionalism in Comparative Politics," *Annual Review of Political Science*, Vol. 2, 1999, pp. 392–396; Kathleen Thelen, "How Institutions Evolve: Insights from Comparative Historical Analysis," in James Mahoney and Dietrich Rueschemeyer, eds., *Comparative Historical Analysis in the Social Sciences*, Cambridge, UK: Cambridge University Press, 2003, pp. 214–216.

mathematically by the Polya Urn model discussed in the appendix.[112] Other mechanisms of reproduction emphasize that a decision may be reproduced because of its impact on underlying power differentials, because it is seen as legitimate or appropriate, or because it becomes increasingly functional. The power mechanism recognizes that decisions can empower certain groups and disadvantage others. In this situation, actors behave rationally, but a decision may be reproduced even if most actors oppose it or prefer an alternative because the elites that benefit from the status quo resist efforts to change it.[113]

The functional mechanism attributes the stable reproduction of a decision to the fact that it is increasingly useful to a larger system. According to this logic, a certain path has increasingly beneficial effects over time that increase its practical utility. Alternatively, legitimacy or the collective belief that a certain path is appropriate can lead to the perpetuation of a decision. The legitimacy-beliefs feedback mechanism suggests that most actors view the current path as appropriate or legitimate and therefore support its reproduction. What differentiates the legitimacy mechanism from increasing returns or power, however, is the fact that actors support—either actively or passively—the reproduction of the current path because of a collective belief that it is the correct course of action, not because of the costs or benefits associated with it.[114]

Fourth, following a critical juncture, there is closure; the range of previously feasible options quickly narrows, limiting future decisions. The decision at a critical juncture initiates one or more of the self-reinforcing processes that strengthen the initial decision; therefore, as time passes, some paths become less likely or completely unattainable.[115] Over time, more and more options are foreclosed until there is little choice but to remain on the same path.

A classic example of path dependence is the QWERTY keyboard.[116] It is far from obvious what led to the dominance of the QWERTY keyboard, which is widely recognized to be a suboptimal layout that hinders typing efficiency. The QWERTY typewriter was designed in 1873 to reduce jamming by arranging common pairs of letters on opposite sides of the keyboard

---

[112] The appendix to this report provides a short introduction to the Polya Urn model and applies it to global posture. For more on the Polya Urn model, see Arthur, 1994, pp. 6–8, and Scott E. Page, "Path Dependence," *Quarterly Journal of Political Science*, Vol. 1, No. 1, 2006, pp. 87–115.

[113] Mahoney, 2000, p. 521; Thelen, 1999, pp. 394–395; Thelen, 2003, pp. 215–216.

[114] Mahoney, 2000, p. 523. This is essentially a constructivist mechanism. For more on constructivism and the importance of inter-subjective beliefs, see Judith Goldstein and Robert O. Keohane, *Ideas and Foreign Policy: Beliefs, Institutions and Political Change*, Ithaca, N.Y.: Cornell University Press, 1993; Peter Katzenstein, *The Culture of National Security, Norms, Identity, and World Politics*, New York: Columbia University Press, 1996; Alexander Wendt, *Social Theory of International Politics*, Cambridge, UK: Cambridge University Press, 1999; Marc Blyth, *Great Transformations: Economic Ideas and Institutional Change in the Twentieth Century*, New York: Cambridge University Press, 2002; and Jeffrey W. Legro, *Rethinking the World: Great Power Strategies and International Order*, Ithaca, N.Y.: Cornell University Press, 2005.

[115] Bennett and Elman, 2006, p. 252; and Mahoney and Schensul, 2006, p. 4.

[116] Paul A. David, "Clio and the Economics of QWERTY," *American Economic Review*, Vol. 75, No. 2, May 1985, pp. 332–337; Boas, 2007, p. 36.

(thereby slowing the typist), and it quickly gained in popularity. By the mid-1890s, a number of typewriter designs had solved the jamming problem and were superior to the QWERTY arrangement. However, QWERTY's initial advantage, which it secured by chance, gave it an edge over its competitors, and between 1895–1905, the main typewriter producers adopted the QWERTY design one by one.

The QWERTY keyboard design came to dominate the marketplace because of the increasing returns mechanism. Because changing to a new keyboard would require retraining all the typists who had learned to use the QWERTY arrangement, the costs of changing were directly proportional to the number of people who had adopted the QWERTY system. Once enough people had accepted the keyboard, it was locked in, even though its inefficient configuration was no longer needed due to mechanical advances in typewriter design.

In short, path dependent processes are situations in which contingent decisions made at a critical juncture end up constraining future choices. Path dependence often helps explain the persistence of institutions or outcomes that seem to be suboptimal or inefficient. Although path dependent processes explain stability where it might not otherwise be expected, they do not preclude change—they just suggest that it will be difficult. Moreover, understanding the specific mechanisms of reproduction at work in a path dependent sequence can help identify the conditions under which change is likely to occur.

Path Dependence and Posture

Typically, the United States establishes an overseas bases in response to a particular threat or to carry out a specific operation. However, even when that mission is completed, U.S. forces often remain at a base, repurposing the facility.[117] It has long been noted that posture is somewhat sticky or difficult to change. For instance, in 1970 the Subcommittee of the Senate Committee on Foreign Relations observed that: "Once an American overseas base is established, it takes on a life of its own. Original missions may be outdated but new missions are developed, not only with the intention of keeping the facility going but often actually to enlarge it."[118] Similarly, a 1979 Congressional Research Service report highlighted the fact that "overseas facilities must be constructed over long periods of time and tend to become self-perpetuating." This fact creates an "inherent tension. . . between the extensive network of U.S. bases overseas and American foreign policy" because the former is fixed, and the latter can change gradually or suddenly.[119]

---

[117] Pettyjohn, 2012, pp. 102–104.

[118] Quoted in Christopher Sandars, *America's Overseas Garrisons: The Leasehold Empire*, New York: Oxford University Press, 2000, p. 16.

[119] Foreign Affairs and National Defense Division of the Congressional Research Service, *United States Foreign Policy Objectives and Overseas Military Installations*, prepared for the Committee on Foreign Relations of the United States Senate, Washington, D.C.: U.S. Government Printing Office, April 1979, p. iii.

Nevertheless, no one has specifically assessed whether and how the concept of path dependence may help explain the durability of the U.S. overseas presence. As we considered how to apply path dependence to posture, it seemed as if a different combination of self-reinforcing processes could be at work at different levels of posture. We, therefore, sought to answer the following questions:

- Why do individual bases persist?
- Why does a U.S. military presence in a particular country persist?
- Why does a U.S. military presence in a particular region persist?
- Why does the United States maintain a global network of bases?

As we considered these questions, we hypothesized that different mechanisms of reproduction were at work to make bases "sticky" at different levels. In any one situation, multiple mechanisms of reproduction can be at work, layered on top of each other, making change even more difficult. For instance, one mechanism of reproduction that holds across all levels in the modern era is the power of the United States.[120] U.S. power, which has varied over time, still has remained relatively high and has provided the foundation for the U.S. overseas military presence. Despite this, the United States has not been able to impose its will on host nations.[121] At times, host nations have expelled U.S. forces or restricted U.S. basing rights. In short, American strength is part of but not the entire explanation of why the United States has bases overseas. Figure 5.2 shows four levels where path dependent processes may be at work in posture.

---

[120] Although clearly U.S. power absolutely and relatively has changed over time.

[121] See Alexander Cooley, *Base Politics: Democratic Change and The U.S. Military Overseas*, Ithaca, N.Y.: Cornell University Press, 2008, and Kent E. Calder, *Embattled Garrisons: Comparative Base Politics and American Globalism*, Princeton, N.J.: Princeton University Press, 2007.

Figure 5.2. Path Dependence at Different Levels of Posture

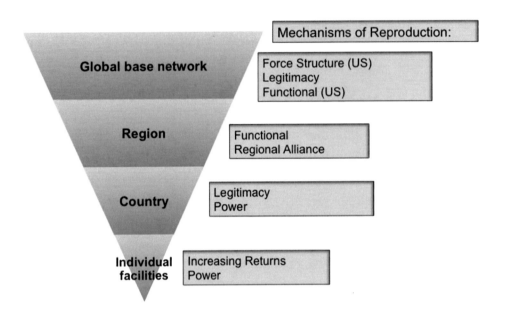

Individual Bases

Path dependence is the most readily apparent at the individual base level. A cursory examination of major USAF bases today reveals that the majority of the bases in Europe and Asia—where the United States has the largest military presence—date back to at least the early 1950s, if not to World War II or before.[122] Moreover, all these legacy bases have been repurposed from their original mission, which was containing communist expansion.[123] Individual bases end up being "sticky" because of several different mechanisms of reproduction, including increasing returns and power.

Increasing Returns

The individual base level seems to best fit the classic economic path dependence logic of increasing returns. Individual bases exhibit three of the four factors—sunk or fixed costs, learning effects, and coordination effects—that are associated with increasing returns.[124]

Air bases are fixed pieces of infrastructure and have large start-up costs; that is, bases are sunk costs. Consequently, there is significant pressure to use existing air bases to the greatest

---

[122] USAF bases that fit this description include Lajes, Mildenhall, Lakenheath, Aviano, Misawa, Kadena, Yokota, Kunsan, Andersen, Incirlik, Ramstein, Spangdahlem, and Osan. See Harry R. Fletcher, *Air Force Bases*, Vol. II: *Air Bases Outside of the United States of America*, Washington, D.C.: Center For Air Force History, United States Air Force, 1993.

[123] For more on repurposing of bases, see Pettyjohn and Vick, 2013, pp. 55–57.

[124] Arthur, 1994, p. 112. The additional factor of adaptation effects does not seem as relevant. See Pierson, 2004, p. 55.

extent possible.[125] As the United States began to expand outside the contiguous U.S. bases after the outbreak of the Korean War, Truman ordered USAF to build new bases only in the absence of an existing USAF or other service base that could be modernized.[126] Because of the sheer number of bases that had been constructed during the war, this meant that USAF typically just improved World War II–era airfields. Overseas, many of the legacy bases were former enemy airfields, such as Misawa Air Base, which had been an Imperial Japanese navy air base.[127] During the Korean War, the absence of airfields capable of supporting USAF aircraft led to an extensive construction program but one that was almost entirely focused on refurbishing and upgrading the infrastructure at existing airfields. In fact, USAF only built one air base—Osan— from scratch.[128]

This tendency to reuse existing airfields is born partially out of a desire to get the maximum return on previous investments and partially for reasons of expediency. In recent times, because of the difficulties of securing land for entirely new air bases, it has also become necessary to repurpose existing facilities or land. Yet this does limit the range of basing options and means that USAF is often held prisoner to the decisions made in earlier eras for different reasons. Unlike early airplanes, USAF's current aircraft require air bases of significant size. As unexploited land has become scarce, land use has become an increasingly contentious issue. Therefore, local communities may resent the U.S. military's control of valuable real estate. For instance, to gain support for new military construction on Guam, DoD has pledged to reduce the land that it controls on the island by 10 percent.[129] Land use has been even more controversial in other countries. In South Korea, for example, local opposition to the presence of Yongsan Garrison—a large U.S. military base—in the heart of Seoul has impelled the United States to move the divisive base. However, securing property for the Yongsan relocation has also been difficult, because the villagers who own the land for the replacement facility resisted the plan to expropriate their farmland.[130]

Coordination effects also provide increasing returns—benefits increase as more people choose the same option.[131] In bases, this takes the form of the expanding facilities, tenants, and infrastructure, which in turns makes the base more capable and attractive. Bases often act like magnets, where a small original presence draws additional capabilities, missions, and tenants.

---

[125] DoD, *Department of Defense Base Realignment Policy*, 1978, p. 13.

[126] Haulman, 2004, p. 55.

[127] Fletcher, 1993, p. 85.

[128] Fletcher, 1993, p. 93.

[129] Tilghman Payne, Commander Joint Region Marianas, briefing to Guam Roundtable participants on Joint Region Marianas, September 5, 2013.

[130] Andrew Yeo, "Local-National Dynamics and Framing in South Korean Anti-Base Movements," *Kasarinlan: Philippine Journal of Third World Studies*, Vol. 21, No. 2, 2006, p. 43.

[131] Pierson, 2004, 24

An example of this phenomenon is the U.S. presence in Camp Lemonnier, Djibouti. When the Combined Joint Task Force–Horn of Africa was established in 2003, it was a USMC operation of approximately 1,500 people. But operations at Camp Lemonnier have grown due to the versatility of the location for both U.S. Africa Command and U.S. Central Command. By 2012, DoD decided that it needed to transition Camp Lemonnier into a permanent compound. As of 2012, this "expeditionary" base had 1,070 structures—36 were permanent; 49 were semipermanent; 356 were relocatable; and 629 were temporary.[132] There were 16 tenant organizations, and Camp Lemonnier's airfield could accommodate 42 aircraft (albeit not parked to DoD standards).[133] Camp Lemonnier was already bursting at the seams, and its steady-state presence was expected to grow to 3,000 U.S. civilian and military personnel. Additionally, there was a new requirement that the base have surge capacity to accommodate an extra 1,500 people.[134]

As the Department of the Navy considered how to meet these demands, it studied whether it should expand its operations at Camp Lemonnier or seek access to a new facility. Ultimately, the Navy concluded that a "new location would be costly and require starting the camp from scratch" but that "this option had the advantages of configuring appropriate land uses from the start, providing room for expansion, and separating incompatible facilities and functions." But these benefits were more than offset by the difficulty of moving, which would have required "purchase or lease of a large area of land and complete development of all infrastructure." Moreover, moving was "the most expensive" option considered and "relied on land/agreements that are not in place."[135] Consequently, the Navy decided to expand and improve the facilities at Camp Lemonnier.

Finally, there are learning effects—the knowledge gained by staying on a certain path. The more that airmen operate at a particular air base, the more knowledge they acquire about the base's layout, the local weather patterns, and the terrain surrounding the facility. This helps make flying operations more efficient and routinized, creating incentives to stay at existing bases.

Power Mechanism

In addition to increasing returns, the power mechanism can encourage the persistence of an individual facility. In particular, three different groups—the local community, the U.S. service that owns the base, and the combatant command (COCOM) area of responsibility that the base resides in—may benefit from its creation and persistence. Over time, local communities develop a level of economic dependency on a base and may loudly protest any effort to remove a facility

---

[132] U.S. Department of the Navy, *Report to Congress on Camp Lemonnier, Djibouti Master Plan*, Washington, D.C.: August, 24, 2012, p. 4–25.

[133] U.S. Department of the Navy, 2012, pp. 2–30.

[134] U.S. Department of the Navy, 2012, pp. 3–7.

[135] U.S. Department of the Navy, 2012, p. 6–20.

that underpins their financial well-being.[136] This is most obvious in the United States, where members of Congress exert great efforts to save bases located in their constituencies, but the symbiotic relationship between towns and U.S. bases develops overseas as well.[137] Since the end of the Cold War, the United States has drawn down its military presence in Germany, causing a significant amount of economic dislocation. For example, the closure of Bitburg Air Base in 1994 had an enduring adverse impact on employment in the surrounding area.[138] Moreover, economic dependency seems to deepen the longer a base has been established. When the Bush administration announced its intention to reduce the U.S. military presence in Germany, which in most instances had been in place since the early 1950s, officials from many German localities visited Washington to lobby against the planned closures.[139] One of the differences between local politics at home and abroad is that the former has the means to influence basing decisions through Congress, while the latter has no direct way of securing their desired outcome. Today, there is concern in the United Kingdom that the plan to close RAF Mildenhall will cost the local economy $331 million a year.[140]

Additionally, two U.S. military parties—the service that owns the base and the COCOM— are vested in overseas installations and may therefore try to frustrate efforts to shut down their bases. The service that owns the base might fight closure because the loss of a large MOB is often associated with a corresponding reduction in force structure. For its part, a COCOM might oppose reductions to its military footprint because they lessen COCOM capability and influence. One of the most effective obstructionist tactics is delaying shutdown in the hopes that the decision will be reversed. Since closing bases takes a long time—years, if not decades—there are ample opportunities for opponents to derail the closure. A prior decision to shutter a facility may be overturned by a new Secretary of Defense or a new administration. Alternatively, circumstances may change so that the base has renewed value or a new purpose is found.

The George W. Bush administration's 2004 announcement that it planned to redeploy two U.S. Army divisions from Germany to the United States illustrates some of these dynamics.[141] In

---

[136] On the other hand, air bases also have significant downsides, including the amount of land that they occupy and the noise and the hazards associated with operating aircraft. Not unreasonably, neighboring communities aim to limit these undesirable side effects. The negative externalities associated with air bases, therefore, can create significant tension between USAF and locals. Calder, 2007, p. 84.

[137] Calder, 2007, pp. 94–95.

[138] Keith B. Cunningham and Andreas Klemmer, *Restructuring the US Military Bases in Germany: Scope, Impacts, and Opportunities*, Bonn, Germany: Bonn International Center for Conversion, Report 4, June 1995, p. 28–30.

[139] Calder, 2007, p. 95. The authors wish to thank RAND colleague Andy Hoehn, who was deeply involved in posture realignment efforts during the George W. Bush administration, for describing the efforts of local officials to prevent base closures in Germany.

[140] "RAF Mildenhall: Business Concerns as the USAF to Leave," British Broadcasting Corporation, January 9, 2015.

[141] This example draws on an unpublished paper: Brian Arakelian and Rich Davison, *Redeployment of Army Forces from Germany: An Analysis of the U.S. Domestic Context*, Washington, D.C.: National Defense University, 2011.

2006, the commander of U.S. European Command (USEUCOM), Gen. James Jones, asked DoD to keep an expanded presence in Germany. Although Rumsfeld had denied Jones's request, the next USEUCOM commander, General Bantz Craddock, petitioned Rumsfeld's successor—Robert Gates—the following year to have the realignment overturned. Due to the 2007 initiative to expand the U.S. Army, Gates postponed the relocation of U.S. Army units from Germany to Texas by several years but did not overturn the earlier decision. In this instance, USEUCOM's pleas only delayed the move; the U.S. Army units left Germany in 2012 and 2013.

For its part, the U.S. Army also initially had concerns about the German realignment plan, fearing that the move would lead to the disestablishment of the two divisions, but these worries were allayed by linking the overseas move to Base Realignment and Closure. By including the contiguous United States stationing of the two divisions in the formal 2005 Base Realignment and Closure process, the Bush administration assuaged U.S. Army concerns that forces would be eliminated.

Country

The reason that the United States maintains a base or bases in a particular country may differ from the reasons that individual facilities tend to endure. At the national level, it appears as if power and legitimacy play a larger role in fostering stability than increasing returns do.

Some countries desire a U.S. military presence because they view it as necessary and appropriate.[142] Many nations have been socialized to believe that a U.S. security guarantee requires forward-stationed U.S. forces. This practice became the norm during the Cold War as the United States sought to convince Moscow and its European allies that it was willing to risk New York or Washington to defend Bonn and Paris. Over time, this has become the standard, and now elites in other nations generally believe that a forward presence is necessary for a U.S. extended deterrent commitment to be credible. As a result, many partners view with dismay the prospect of a U.S. military drawdown and attempt to prevent this from happening. For instance, allies in other parts of the world have voiced concerns that DoD's fiscal constraints and its strategy of rebalancing toward the Asia-Pacific will lead the U.S. military to abandon their regions.

Partners in the Persian Gulf, in particular, viewed the rebalance as a sign that the United States was disengaging from the Middle East.[143] These concerns were heighted by the prospect of a nuclear deal with Iran. In an effort to assuage the Gulf states' fears, Secretary of Defense

---

[142] Yeo uses host-nation elite security consensus as the key variable to explain whether antibase movements succeed or fail. In countries where there is an elite security consensus in favor of a U.S. military presence, the government will reject antibase activists' demands, leading to the perpetuation of U.S. bases even when they are broadly unpopular (Andrew Yeo, *Activists, Alliances, and Anti-U.S. Base Protests,* Cambridge, UK: Cambridge University Press, 2011, pp. 14–17, 21–24).

[143] Bilal Y. Saab and Barry Pavel, *Artful Balance: Future U.S. Defense Strategy and Posture in the Gulf,* Washington, D.C.: Atlantic Council, March 2015, p. 3.

Chuck Hagel promised in December 2013 that "The Department of Defense will continue to maintain a strong military posture in the Gulf region" and that "DoD will not make any adjustments to its forces in the region or to its military planning as a result of the interim agreement with Iran." Going even further, Hagel took the unusual step of enumerating the number and types of U.S. troops and capabilities in the region.[144] Thus far, these fears have been unfounded; the emergence of ISIS has meant that the United States has actually expanded its military footprint in the region.

Alternatively, enduring partner nations view U.S. forward-based forces as legitimate because of an elite security consensus that the U.S. military is a force for good that helps promote stability and prosperity.[145] This perspective has become entrenched after many decades of hosting U.S. forces for much more parochial reasons tied to partners' own national security. The view that a U.S. military presence is appropriate helps perpetuate U.S. bases in enduring partner nations, such as the United Kingdom, Germany, Italy, and Japan.

In contrast, powerful elites in a host nation may support and benefit from the presence of U.S. forces in country and therefore push for their persistence. These benefits may enable the host country to focus on economic growth while relying on the United States for security or may enrich the privileged few.[146] For instance, at Manas Air Base in Kyrgyzstan, the United States awarded fuel contracts to companies run by President Askar Akayev's son and son-in-law. This encouraged Akayev to maintain the U.S. presence, even if the Kyrgyzstani public disliked it. When a popular revolution ousted Akayev in 2005, his successor, President Kurmanbek Bakiyev, and Bakiyev's son expropriated the two fuel subcontractors and continued to receive U.S. contracts. Despite the change in leaders, the Kyrgyzstani elite remained vested in the U.S. base. Yet the perceived corruption and fraud at Manas contributed to the discontent that fueled popular revolutions in Kyrgyzstan in 2005 and 2010 and embittered the Kyrgyzstani population toward the United States, leading ultimately to the eviction of U.S. forces in 2014 after Bakiyev was deposed.[147]

The power feedback mechanism may help to explain why regime change—when a state experiences a coup or a revolution—usually results in the expulsion of U.S. forces.[148] New elites entering office might not necessarily benefit from the U.S. military presence; in fact, their

---

[144] Charles Hagel, "Remarks by Secretary Hagel at the Manama Dialogue from Manama, Bahrain," December 7, 2013.

[145] Pettyjohn and Vick, 2013, pp. 55–56.

[146] Cooley and Nexon, 2013, pp. 1039–1040.

[147] U.S. House of Representatives, *Mystery at Manas: Strategic Blind Spots in the Department of Defense's Fuel Contracts in Kyrgyzstan*, Washington, D.C.: Report of the Majority Staff, Subcommittee on National Security and Foreign Affairs, Committee on Oversight and Government, December 2010, pp. 13–14; Cooley and Nexon, pp. 1042–1043; and Alexander Cooley, *Great Games, Local Rules: The New Great Power Contest in Central Asia*, Oxford, UK: Oxford University Press, 2012, pp. 120–130.

[148] Cooley, 2008, pp. 249–253; Calder, 2007, p. 228.

credibility often rests on opposing the presence of foreign forces. If they do acquire benefits—private or public—from U.S. bases, as occurred in Kyrgyzstan in 2005, they are likely to support a continued U.S. presence, but if they do not profit from the situation, they are more likely to push for a U.S. presence to be reduced or eliminated.

Region

It was less clear whether there were path dependent processes at work at the regional level. After all, the U.S. military presence in a region is really just an aggregation of individual bases in particular countries. Nevertheless, it seemed plausible that there are independent functional and the legitimacy feedback mechanisms at work at the regional level.

The functional logic is straightforward: The United States creates bases in a region for one reason, for example, containing the Soviet Union, but these bases acquire other useful purposes, such as projecting military power globally or into other regions. This was certainly the case during the Cold War. The primary mission of U.S. bases in Western Europe was to deter Soviet aggression. Nevertheless, by the late 1950s, the United States had discovered that these bases also enabled U.S. forces to deploy to and operate in other regions. By the 1970s, the issue of using European bases for out-of-area operations had become contentious because many NATO allies were reluctant to allow the United States to use their bases for any mission other than NATO defense. After the Cold War ended, the primary function of U.S. bases in Europe shifted from deterrence and the defense of Europe against a Soviet invasion to supporting expeditionary operations in other locations.[149] U.S. bases in Europe remained after the end of the Cold War because they enabled U.S. forces to carry out operations in other regions.

In contrast, the legitimacy mechanism holds that a U.S. presence in a region is largely due to the fact that countries within that region view it as appropriate. It would seem that this is the case in Europe, although U.S. bases in Europe are also intimately tied to the persistence of the NATO alliance. In other regions, such as the Middle East and Africa, U.S. forces are not generally as welcome because they often are seen as undermining the host nation's independence.

To explore whether there was evidence of path dependence, we looked at the number of major USAF bases by region. A glance at the raw number of bases shows that there is considerable variation (see Figure 5.3), which in turn suggests that path dependence may not apply to the regional level.[150]

---

[149] Pettyjohn, 2012, Chapter Ten. Support included flying combat operations from Aviano Air Base in Operations Deliberate Force and Allied Force and from Incirlik Air Base during Operation Northern Watch.

[150] See the appendix for an additional analysis using the Polya Urn model.

Figure 5.3. Number of Major USAF Air Bases by Region, 1953–2011

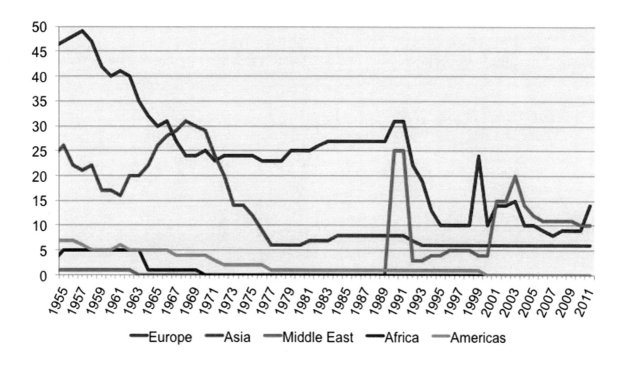

Nevertheless, we examined other metrics—the change in the proportion of bases by year by region and the percentage of bases by region—that better capture relative, rather than absolute, changes in USAF posture. Given that there were few bases in Africa and Latin America, we examined three primary regions: Europe, Asia, and the Middle East. The change in total number of USAF bases from the previous year, as displayed in Figure 5.4, understates changes that occurred gradually and does not depict the larger downward trend.[151] Nevertheless, this metric helped us identify break points—moments when there may have been an exogenous shock or a critical juncture. Thinking of USAF's posture as a punctuated equilibrium model suggests that it is more appropriate to expect greater stasis within one of these periods instead of across the entire time examined. For instance, during the Cold War (1953–1989), the change never exceeded 10 percent, and the largest perturbations occurred during the U.S. wars in Southeast Asia. Therefore, for 1953–1989, there is some support for the notion of path dependence. There was clearly a large shock in 1990, as the number of bases in Europe plummeted and the number of bases in the Middle East spiked due to the Gulf War. As a period, the 1990s were quite volatile, with several spikes in Europe and the rapid buildup in the Middle East, followed by an

---

[151] Care should be taken when interpreting this graphic. It is not intended to imply causality or a negative correlation between percentage changes across the three regions. The graphic displays changes for each region as a percentage of all the bases (in Europe, Asia, and the Middle East). Effectively, this means shifts in one region can and will impact the total percentage change in other regions. An alternative approach would rebase percentage changes in each region strictly according to the number of bases in the same region in the prior year. Unfortunately, given the number of zeros in the data for the Middle East, this is not practical from a graphing perspective.

61

equally rapid drawdown, and then a gradual yearly increase. Another shock occurred with the 9/11 attacks on the United States; once again, the number of bases in the Middle East increased significantly. After that period, things settled down for the next decade.

Figure 5.4. Change in the Total Number of USAF Major Air Bases from the Previous Year

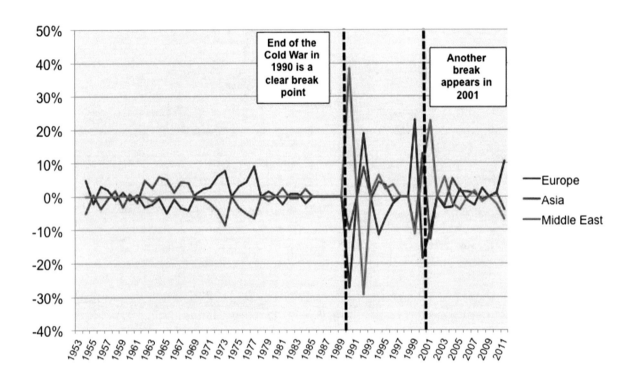

It seemed that the proportion of bases, rather than the absolute numbers, might be a better indicator of path dependence. The proportion of bases by region provided further limited support for the idea of path dependence within shorter periods, especially during the Cold War (see Figure 5.5). For instance, looking only at the Cold War data, Europe had the largest percentage of USAF bases except for the years of the Vietnam War. The proportion of bases by region confirms that 1990 was a shock—or a critical juncture—as two new trends were established: There was a significant reduction in USAF's presence in Europe and an increase in the Middle East. There were two additional upticks in Europe with the operations in Bosnia and Kosovo, but neither of these resulted in enduring changes to USAF's posture. Another break point occurred in 2001, leading to significant growth in the number of USAF bases in the Middle East and continued decline in Europe (Asia remained relatively flat).

62

Figure 5.5. Proportion of Bases by Region, 1953–2011

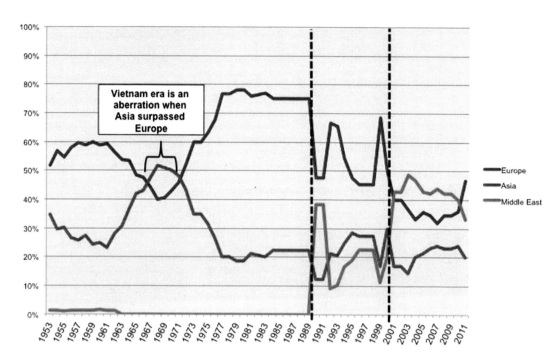

It is necessary to look at several different metrics to get a complete picture of USAF posture at a regional level. In short, there is limited evidence that path dependence holds at the regional level, particularly during the Cold War. During this time, the network of USAF bases was nearly constantly revised, but during shorter periods, there has been a good bit of stasis. For instance, Europe remained the region with the largest percentage of USAF bases until 2001; even after 2001, it had more USAF bases than Asia, which had flatlined by the 1980s (see Figure 5.3). While fluctuations on the margins occurred continuously, very large shifts seemed to require an exogenous shock. This makes sense. When a situation is in flux due to crises, it is easier to overcome bureaucratic, host-nation, and domestic political resistance to change.

Global Base Network

Today the U.S. network of overseas bases is often taken for granted by American officials and other nations, which expect the U.S. military to quickly respond to crises, disasters, and threats. But for much of its history, the United States did not have a robust overseas military presence. Once the United States established an extensive global network of bases after World War II, this presence became entrenched and does not appear likely to disappear anytime soon. The history of how this came to be exhibits characteristics of path dependence: initial openness, chance, a critical juncture, and lock-in.[152]

Before World War II, the United States had a very limited military presence overseas, mainly on U.S. colonial possessions to defend U.S. commercial interests. World War II—in particular,

---

[152] This section is drawn from Pettyjohn, 2012.

the attack on Pearl Harbor—prompted a shift in the thinking of U.S. defense planners. Previously, overseas bases were seen as unnecessary and likely only to draw the United States into conflicts that it had little stake in. But World War II reversed that mindset. Spurred by a feeling of deep and enduring vulnerability, U.S. defense officials concluded that the nation's security could only be guaranteed by fighting and defending forward. As a result, the JCS developed a series of plans for a vast network of mainly air and naval bases to extend the U.S. defensive perimeter. Yet after the war had ended, most nations were unwilling to permit the United States to establish a permanent military presence, and Congress refused to finance the ambitious basing plans. As tensions with the Soviet Union increased, a few nations relented, and Congress loosened the purse strings, but a sea change only occurred after North Korea invaded South Korea, prompting fears of communist aggression worldwide.

In the early 1950s, the United States established hundreds of large bases overseas, mainly in Western Europe and East Asia, to deter communism. This containment posture of consolidated defense in depth differed significantly from the one initially envisioned by the JCS during World War II. Instead of a network of relatively austere air and naval bases, the United States established a large permanent footprint in Western Europe that included ground forces to defend the Federal Republic of Germany. As a result of this choice, U.S. allies and partners came to expect that an American security guarantee would be backed up by forward-based forces. At the same time, U.S. leaders discovered that a worldwide network of bases enabled the United States to project power globally. These two factors helped make the U.S. base network "sticky."

The Cold War came to an end with the revolutions in Eastern Europe in 1989 and the collapse of the Soviet Union in 1991. Given the improved security situation, the United States opted to significantly cut its defense expenditures by drawing down the size of its military and reducing the number of U.S. bases in Europe. Despite the absence of a peer competitor, the United States ultimately retained a small portion of its bases and forces in Europe, as well as much of its military presence in Asia. It did so because it found that these were useful platforms for power projection and that they also helped maintain relationships and interoperability with host nations and allies.

At the network level, two mechanisms of reproduction seem to be the most important: legitimacy and functionality. As explained in the country discussion, many countries have come to accept a U.S. military presence as not only necessary but appropriate. Bases abroad enable the United States to provide important collective goods—such as freedom of the commons—and can help promote stability by deterring aggression and assuring allies. Functionally, U.S. bases enable the United States to carry out military operations almost anywhere on the globe.

These feedback mechanisms are reinforced by a related but separate factor: USAF force structure. During the Cold War, forces were acquired to defend the European central front from a Soviet invasion. This necessitated a large force of USAF fighter aircraft to conduct critical missions, such as air defense and close support of ground forces. Only fighters could accomplish these missions, but fighter aircraft of the early Cold War period were short ranged and not air refuelable, necessitating a large network of forward air bases. Although long-range aircraft made important contributions in interdiction, strategic attack, and nuclear deterrent missions, short-range fighters made up the bulk of the conventional warfighting force. Even long-range aircraft of the era required overseas bases. With the advent of the B-52 and a large aerial refueling fleet, the nuclear deterrent mission largely moved back to bases in the United States, and fighters could operate from more-distant bases with air refueling. Despite these changes, USAF still required a large number of forward bases to generate the number of combat sorties needed to win a major war in Europe.

The existence of a large network of high-quality forward bases developed in the early Cold War years may have created a modest degree of path dependence in USAF aircraft acquisition over subsequent years and decades. Although combat requirements, physics, and technological limitations were the primary drivers of USAF force structure, if the United States had not possessed such an extensive network of overseas airfields, the relative balance of shorter- and longer-range platforms in USAF inventory might have shifted somewhat toward longer-range aircraft.[153] Evidence in support of this can be found in current analyses regarding air operations in antiaccess or area denial environments, where forward bases are expected to be under heavy attack (the functional equivalent of a smaller network of forward bases—at least during some periods of the conflict). Long-range strike platforms are expected to play a larger role, and some analysts argue for a shift in force structure toward a longer-range force.[154]

Path Dependence in the Post–Cold War Era

As the analysis above revealed, USAF posture was more path dependent during the Cold War than in the post–Cold War era. During the Cold War there was greater stability in terms of both the balance of power and the focus of U.S. defense planning. In the post–Cold War era, while the United States has remained the hegemon (although its ascendency may be eroding), it has faced a diverse array of security challenges, such as regional adversaries (including Iraq, Iran, and North Korea), terrorist organizations, and gray zone aggression by near-peer competitors.[155] To counter these geographically dispersed and varied challenges, the United

---

[153] Shifts in the relative power of the fighter and bomber communities in USAF may have contributed somewhat to this as well. See Mike Worden, *Rise of the Fighter Generals: The Problem of Air Force Leadership, 1945–1982*, Maxwell AFB, Ala.: Air University Press, March 1998.

[154] For a discussion of the evolving threat to forward air bases, see Alan J. Vick, *Air Base Attacks and Defensive Counters: Historical Lessons and Future Challenges*, Santa Monica, Calif.: RAND Corporation, RR-968-AF, 2015a. Assessments arguing that USAF force structure is overly weighted toward shorter range aircraft include Peter Garretson, "A Range-Balanced Force: An Alternative Force Adapted to New Defense Priorities," *Air and Space Power Journal*, May–June 2013, pp. 4–29; Mark A. Gunzinger and David A. Deptula, *Toward a Balanced Combat Air Force*, Washington, D.C.: Center for Strategic and Budgetary Assessments, 2014; and John Stillion, *Trends in Air-to-Air Combat: Implications for Future Air Superiority*, Washington, D.C.: Center for Strategic and Budgetary Assessments, 2015.

[155] Fareed Zakaria, *The Post American-World: Release 2.0*, New York: W.W. Norton & Company, 2011.

States has made some fairly significant revisions to its posture. It seems more likely that the future will resemble the past few decades, rather than the Cold War; therefore, we should anticipate more changes to U.S. posture.

Because the international environment has been characterized by greater uncertainty, multiple U.S. administrations have emphasized that DoD is transitioning from large fixed locations to a more-agile basing network with a greater number of "lighter" facilities. These so-called "lily pad" locations—forward operating sites (FOSs) and cooperative security locations (CSLs)—are smaller, more-austere facilities that U.S. troops periodically deploy to without their families, enabling DoD to easily make adjustments to its posture to meet demands.[156] Light facilities are intended to prevent path dependence from locking the United States into using a particular base, country, or region. Yet the question remains whether they are likely to succeed at this task.

Theoretically, smaller, more-scalable facilities could weaken or even undermine several of the path dependent mechanisms of reproduction that help produce stasis. First, if the United States does not make significant investments in partner nation facilities, there will be fewer sunk costs and less pressure to maintain or reuse an existing base. In part, this idea underpins the strategy of creating "places not bases."[157] Places are partner-nation facilities that U.S. forces occasionally fall in on and operate from. They allow the United States to quickly and easily expand its basing network as new requirements arise and contract it as old requirements fade away. If the United States does not have to build entirely new bases or spend its money on improving these locations, it should be more willing to walk away—at least in theory. Furthermore, smaller bases should offer fewer coordination effects. Because only a handful of U.S. personnel or activities are permanently located at a specific facility, there are few benefits to expanding at that particular location.

Second, lighter facilities should weaken the impact of the power mechanism at the individual base level. While FOSs and CSLs could still empower host-nation leaders, they are less likely to significantly boost the local economy because of the sporadic nature of the U.S. military presence. Similarly, because these lighter facilities do not have permanently assigned large U.S. military units, the services are less likely to object to closing them because doing so is not likely to result in cuts to force structures.

Finally, lighter-footprint bases could unintentionally undermine the legitimacy of U.S. bases in a country, region, or even at the global network level because of their secretive nature and their association with remotely piloted aircraft (RPA) strikes. The locations of many CSLs are kept hidden from the general public because of the sensitive nature of the missions at these bases or because host nations do not want to publicize the existence of U.S. military bases. Moreover, DoD has deployed RPAs to a good number of these smaller facilities, and RPAs are

---

[156] DoD, 2004, p. 10.

[157] Pacific Air Forces, *Command Strategy, Projecting Airpower in the Pacific*, Hickam AFB, Hawaii, 2014, p. 7.

controversial platforms, given their use for lethal strikes. A July 2014 Pew poll found that, in 39 of the 44 countries surveyed, majorities or pluralities of the population opposed U.S. RPA strikes.[158] Given widespread misgivings about RPAs, their association with lighter-footprint bases may provoke opposition to these types of bases and increase the probability that U.S. forces are expelled. Because popular opinion is quite against these attacks and because there is considerable disinformation about RPAs and fear of their ties to FOSs and CSLs, such attacks could undercut support and eventually change views about the appropriateness of U.S. bases.

At the network level, U.S. bases face an additional challenge to their legitimacy. Despite the fact that the U.S. network of bases currently seems quite resilient, the international environment is changing in ways that could make it more difficult for the United States to secure and maintain access to bases abroad. In particular, the growing importance of nationalism and popular sensitivity regarding sovereignty have generated greater public opposition to U.S. military presence, and the spread of democracy have compelled host-nation governments to take public opinion into account.[159] Moreover, these challenges have arguably become even greater in the post–Cold War era due to the proliferation of information and communications technologies, which have helped publicize America's global military presence and mobilize local, national, and transnational opposition networks.[160] Therefore, norms could shift in ways that might undermine the legitimacy of the U.S. overseas military presence.

In theory, a "places not bases" approach to posture should reduce the effects of path dependence, making it easier to expand, contract, or realign the global network of U.S. bases. In practice, however, the evidence to date is mixed. Path dependent mechanisms of reproduction appear to be at work in many locations that are considered to be FOSs and CSLs. In part, this is due to the fact that DoD often has to finance improvements to its partners' existing military facilities so that they can support U.S. forces. Occasionally, allies will share the costs for base upgrades. For instance, Australia and the United States are both underwriting the expansion of Robertson Barracks in Darwin so that it can accommodate the 2,500 U.S. Marines who will rotate there for six months of every year.[161] In general, however, most of the countries where the United States wants to establish light-footprint bases do not have the resources to subsidize base enhancements, leaving the United States to foot the bill. In 2009, for instance, DoD spent $100

---

[158] Only in three nations—Israel, the United States, and Kenya—did the majority of the population support this policy (Pew Research Center, "Global Opposition to U.S. Surveillance and Drones, but Limited Harm to America's Image," July 2014, p. 5).

[159] Cooley and Spruyt, 2009, p. 1.

[160] Andrew Yeo, "Not in Anyone's Back Yard: The Emergence and Identity of Transnational Anti-Base Network," *International Studies Quarterly*, Vol. 53, No. 3, September 2009, pp. 571–594.

[161] Rob Taylor, "Australia Embraces Marine Presence in Darwin," *Wall Street Journal*, August 14, 2014. Additionally, many wealthy Gulf nations have subsidized U.S. bases in their nations. See U.S. Senate, *The Gulf Security Architecture: Partnership with the Gulf Cooperation Council,* majority staff report prepared for the Committee on Foreign Relations, Washington, D.C.: U.S. Government Printing Office, 2012.

million to develop the infrastructure at Romanian and Bulgarian bases.[162] While this is a relatively modest sum for the Pentagon, once an investment has been made there is an incentive to make the most of the facility, leading to further development. Mihail Kogalniceanu Air Base (MKAB) in Romania, for instance, was initially supposed to have only 100 permanently assigned U.S. military personnel, but after the 2014 eviction from Manas, DoD continues to build up MKAB to make it a major transit hub to and from Afghanistan.[163] Similarly, the Navy was planning to spend roughly $1.4 billion to make improvements to Camp Lemonnier—an FOS—in 2012.[164]

Air bases require minimum infrastructure (a runway and parking spaces), which often needs to be upgraded to support large USAF jets, but all the services end up needing to make investments in bases. For instance, the U.S. Army has asked for $70 million and $58 million in fiscal years 2016 and 2017, respectively, to improve ranges and training sites in Eastern and Central Europe as a part of the European Reassurance Initiative.[165]

As a result of these dynamics, many bases that DoD considers to be FOSs and CSLs actually greatly resemble MOBs, with extensive infrastructure and a continuous U.S. military presence. For instance, the United States reportedly plans to keep 13,500 troops at three major bases in Kuwait, while 7,500 troops are stationed in Qatar, mainly at Al-Udeid Air Base.[166] The main difference is that U.S. troops are deployed on relatively short rotations to lily-pad bases, instead of having U.S. units permanently based overseas (e.g., the 48th Fighter Wing at RAF Lakenheath).

In sum, lighter-footprint bases should, in theory, weaken some of the path dependent mechanisms that encourage individual bases to persist. But in practice, it appears that DoD investments in and fairly large continuous U.S. presence at many FOSs and CSLs may be making change difficult because of the increasing returns mechanism. In contrast, it seems that path dependence at the network, regional, or country levels could be undermined by a change in beliefs about the appropriateness of U.S. bases. In particular, strengthening sovereignty and growing nationalism could weaken collective beliefs about the appropriateness of U.S. bases. Moreover, U.S. bases' association with drone strikes, which are deeply unpopular in most of the world, could undermine the legitimacy of U.S. bases.

---

[162] Seth Robson, "New Bases in Bulgaria, Romania Cost the US Over $100M," *Stars and Stripes,* October 17, 2009.

[163] U.S. Embassy Bucharest, "U.S. Military Engagements to Romania," May 27, 2008; Joe Gould, "In Europe, US Army Official Favors Host-Nation Basing," *Defense News*, March 9, 2015. In 2014, MKAB had approximately 500 personnel and processed up to ten flights a day. The U.S. portion of the base spanned 850 acres. U.S. Army Europe, "21st TSC MK Air Base Fact Sheet," November 3, 2014.

[164] U.S. Department of the Navy, 2012, p. 9.

[165] Office of the Under Secretary of Defense (Comptroller), *European Reassurance Initiative Department of Defense Budget Fiscal Year (FY) 2016,* February 2015, p. 9, Office of the Under Secretary of Defense (Comptroller), *European Reassurance Initiative Department of Defense Budget Fiscal Year (FY) 2017*, February 2016, p. 7.

[166] U.S. Senate, 2012, pp. 12–15.

Conclusions

The concept of path dependence helps explain why so many of USAF's bases are more than 70 years old and why changing USAF posture—either adding, removing, or realigning bases—is difficult. Various path dependent processes work at multiple levels of posture, reinforcing each other and contributing to overall stickiness. As a result of path dependence at the individual base level—which is characterized by increasing returns and power feedback mechanisms—the United States can be tied to a suboptimal base. A facility, created at a different time for a different purpose, may not be well suited for its current mission. Alternatively, the United States may be stuck at a base that is strongly opposed by the local population. Nevertheless, it is clear that path dependence by itself does not determine the shape of USAF's posture. Realignment on the margins occurs nearly continuously, but at the regional level, it appears that an exogenous shock is needed to precipitate sudden large shifts.

Path dependence is often considered to be a bad thing, a process that helps explain inferior or inefficient outcomes. This certainly can be true at the individual base level, but this is not the entire story. From another perspective—the base network level—path dependence is actually beneficial. Without path dependence, the United States might have given up bases more readily, only to regret it as an evolving security environment created new demands for overseas presence. Thus, somewhat counterintuitively, path dependence has helped the United States maintain a global network of overseas bases, which provides it with unique advantages.

Path dependence is largely the product of policy successes (i.e., choices that create increasing returns and other benefits). In contrast, contingent events have led to poor choices and policy failures. When that happens, U.S. policymakers, elites, and the public may react to the failure through a process of "path avoidance," in which particular military missions or operations in problematic terrain types, regions, or against difficult classes of opponents are avoided as policy options. The next chapter explains this concept more fully, explores historical examples, and considers the implications of path avoidance for USAF posture.

6. Path Aversion As a Constraint on Posture

Introduction

As noted in the previous chapter, contingent events do not always set in train path dependent processes. In some cases, policy choices in response to the event go badly and lead to a countervailing process that we call path aversion. Path aversion describes situations where a policy failure is so traumatizing that it leads the president, Congress, policy elites, the military, or the public to resist military operations of certain types or in particular regions. This chapter offers a theoretical and historical discussion of path aversion and its implications for USAF global posture. Path aversion is the fourth and final variable in our model of U.S. global posture (along with strategy, contingent events, and path dependence). We argue that these four driving mechanisms explain both change and stasis in U.S. global posture over extended periods.

The policy failures precipitating path aversion act as critical junctures that (during a war) can lead to reductions in overseas posture as forces are withdrawn prior to the completion of U.S. objectives (e.g., Vietnam and Iraq). In the postconflict period (particularly during the early years), path aversion associated with the policy failure can lead to changes in service doctrine and force structure, congressional limits on executive freedoms, and interventions avoided because the setting appears similar to that of the previous failure.

When the United States avoids or constrains interventions because of path aversion, it inevitably forgoes global posture opportunities. This negative effect is difficult to prove because the avoided or constrained intervention creates no (or little) need for new overseas posture; thus, there is no documentary trail of policy studies, memoranda, or deliberations regarding the relative merits of various bases. That is, no such studies or discussions occur in the absence of a need to support deployed U.S. forces.

A simple example of this is the highly constrained U.S. support provided to the government of El Salvador during its civil war in the 1980s. If the Vietnam War memory had not been so fresh, the United States likely would have followed its previous practice of using U.S. combat forces to maintain friendly governments in power. The deployment of U.S. ground and air forces scaled to defeat Farabundo Marti National Liberation Front (FMLN) insurgents would have swamped the limited El Salvadoran military base network, requiring the construction of new bases and expansion of existing infrastructure. Support bases might also have been created or expanded in nearby countries, particularly Honduras. Although it cannot be proven, the historical experience with path dependence described in the previous chapter suggests that USAF would very likely still possess at least one air base in El Salvador today if not for path aversion following the Vietnam experience.

How Does Path Aversion Relate to Path Dependence?

Path aversion is intended to describe forces that disrupt continuity in policy, whereas path dependence is helpful as a concept that explains forces resistant to change in policy. Historically rare, path aversion appears to occur only in reaction to exceptionally painful, prolonged, and obvious policy failures. Although path aversion is a countervailing force to path dependence, it is neither its theoretical nor empirical equal.

As discussed in the previous chapter, the concept of path dependence originated in the field of economics in the mid-1980s. It offered an explanation for the phenomenon of increasing returns, something that macroeconomic theory once saw as impossible. The concept took hold and found a following among political scientists as well. A large, diverse body of scholarship provides a strong theoretical, mathematical, and empirical foundation for path dependence, including many detailed case studies of path dependent processes. Path aversion has no comparable pedigree, having been coined by our study team in 2015.[167] Although the idea was contained in the idea of the "Vietnam Syndrome," we are not aware of any theoretical or conceptual treatment akin to that for path dependence. The international relations literature does contain many works on the process of learning that follows foreign policy failures, but the emphasis is on alternative strategies rather than avoidance of particular challenges. For example, Robert Jervis, in his classic work, *Perception and Misperception in International Politics*, describes how policy makers use historical analogies in the selection of strategies:

> the explicit decision to avoid strategies that have failed is coupled with an increased readiness to perceive a wide range of situations as resembling the ones that have previously caused the state the most trouble. Once actors believe that they are in a similar situation, they naturally and often wisely adopt a policy different from the one that recently led to disaster.[168]

Path aversion could be viewed as a particular type of reasoning by analogy that is either wise or dogmatic, depending on how universal one believes the lessons are from a policy failure.[169] In path aversion, the lessons learned are not primarily about the specific shortcomings of the failed strategy but rather about the choice to intervene in certain types of situations (defined by region, terrain, type of conflict) that are deemed so problematic that they must be avoided outright.

In this chapter, we will provide a short theoretical introduction to path aversion, primarily so the reader can contrast it with path dependence. Most of the discussion, however, is focused on

---

[167] We initially described it as "path rejection," but reviewer Andrew Yeo noted that it has not been sufficiently powerful to fully reject particular policy paths, suggesting "path aversion" instead.

[168] Robert Jervis, *Perception and Misperception in International Politics,* Princeton, N.J.: Princeton University Press, 1976, p. 276.

[169] For more on the use of historical analogies in foreign policy deliberations, see Richard E. Neustadt and Ernest R. May, *Thinking in Time: The Uses of History for Decision Makers,* New York: Free Press, 1986, and Yuen Foong Khong, *Analogies at War: Korea, Munich, Dien Bien Phu and the Vietnam Decisions of 1965*, Princeton, N.J.: Princeton University Press, 1992.

historical cases of path aversion. These cases all demonstrate two striking aspects of path aversion: (1) its enduring power as an idea and (2) its fleeting power as a constraint on interventions abroad.

To better understand path aversion as a concept, consider a pure version that we will call path rejection. In a path rejecting world, traumatic policy failures would lead to complete avoidance of similar military interventions for extended periods of time, measured in years or decades. The path rejection would exclude all interventions in situations characterized by the same type of conflict, terrain, or region. For example, all interventions in civil conflicts, all jungle warfare, or all operations in the Middle East would be dismissed out of hand as unacceptably risky or doomed to failure following a failure with any of those characteristics. In such a world, a policy failure would act as a critical juncture to a new world state in which certain classes of military operations were no longer policy options considered or available to the U.S. government. Although the "Vietnam Syndrome" is sometimes discussed in these terms, its real effects were much weaker. Indeed, we know of no historical cases in which path rejection occurred in this pure form. Table 6.1 compares and contrasts the key characteristics of path dependence with path aversion as experienced historically.

Table 6.1. Characteristics of Path Dependence and Path Aversion

Path Dependence	Path Aversion
Structural	Attitudinal
Stronger over time (via increasing returns)	Weaker over time (as memory dims)
Enduring constraint	Fleeting constraint

The first difference between the concepts is that path dependence is a structural phenomenon, while path aversion is attitudinal.[170] Path dependence happens when early decisions get locked in and reinforced via a self-reproducing process. It does not require conscious intervention to perpetuate. In contrast, path aversion holds only as long as foreign policy actors maintain a strongly negative attitude toward certain types of military operations. This leads to the second difference: the power of these processes over time. Path dependence grows stronger over time via increasing returns and related reinforcing mechanisms. In contrast, path aversion weakens over time as memory dims about the traumatic event and as new challenges call for U.S. action. Over time, the past failure becomes increasingly contextualized as a unique event that is not predictive of failure in a prospective intervention. Put differently, path aversion is a mental framework for understanding past policy decisions. As such, it is an inherently political construct and therefore subject to strongly divergent interpretations.

---

[170] The authors wish to thank Jason Vick for sharing this insight.

As will be discussed in the section on the Vietnam Syndrome, there was no uniform reaction to the U.S. failure in Vietnam. On the political left, there was strong path aversion toward future interventions in civil conflicts. Many defense professionals developed a view that land wars in Asia were problematic and that jungle warfare limited the utility of advanced American technology. Some prominent military officers and conservative revisionists concluded that the war was lost because President Lyndon Johnson micromanaged military leaders, denying them the freedom of action necessary to achieve a decisive victory. Therefore, the problem was not counterinsurgency (COIN), jungle warfare, or Asia but rather bad strategy. Their path aversion was manifested in a "Let's not enter another war unless we are willing to fight it with fewer constraints" attitude. The public at large embraced a view that had elements of all these perspectives, but the bottom line was clearly "Bring the troops home and keep them out of such conflicts."

Having distinguished between path dependence and path aversion, we will now turn to a more detailed consideration of path aversion, first conceptually, then empirically.

Mechanisms and Dimensions of Path Aversion

The path aversion concept does not argue that a military failure leads to a total cessation of U.S. military activities abroad (that has not happened) but rather that certain types of activities (e.g., COIN operations with U.S. ground forces) are avoided and in the process other military activities (e.g., U.S. ground forces acting as advisers) are constrained due to strong risk aversion by some combination of national leaders, policy elites, and the broader public. Whether and how public opinion drives or follows policy decisions remains a matter of debate among political scientists and is a topic well beyond the scope of this report. For that reason, we will discuss evidence of path aversion in all three groups but will not make any claims about specific causal pathways among them (e.g., that loss of public support for an operation causes path aversion in political leaders).[171] That said, there are clear cases in which political leaders spoke and acted like they believed that the loss of public support undermined the country's ability to wage war successfully. For example, in his famous November 3, 1969, "Silent Majority" speech, President Richard Nixon noted that "the war was causing deep division at home" and that an immediate withdrawal of all American forces "would have been a popular and easy course." His speech sought to build support for an alternative policy (Vietnamization) that would gradually shift the

---

[171] The most prominent recent scholarship argues that the opinions of average citizens have little direct impact on policy formulation. See Martin Gilens and Benjamin I. Page, "Testing Theories of American Politics: Elites, Interest Groups, and Average Citizens," *Perspectives on Politics*, Vol. 12, No. 3, September 2014, pp. 564–581. There also is a body of scholarship (elite cueing) that sees public opinion following the views of partisan leaders. See John R. Zaller, *The Nature and Origins of Mass Opinion*, Cambridge, UK: Cambridge University Press, 1992.

burden of combat operations to the Republic of Vietnam's forces. As they became more capable, U.S. forces would be withdrawn in parallel.[172]

Path aversion may occur on any of several dimensions, including the location of operation, type of mission, type of force, focus of military doctrine, strategy choice, force structure mix, or U.S. presence in a region. Path aversion is typically manifested in one (or more) ways: (1) force withdrawal from an ongoing conflict, (2) institutional shifts in focus away from the problem mission or region, (3) institutional initiatives to restrict Presidential autonomy to use force (such as the 1973 War Powers Act), and (4) a reluctance to use force under similar circumstances.

First, during a long, difficult conflict, U.S. leaders may conclude either that their objectives will never be met or that the cost is simply too high and therefore seek to create conditions that would allow the United States to extricate its forces from the conflict and perhaps even the region. The Vietnamization program just mentioned and the withdrawal of U.S. forces in Iraq (2007–2011) are two prominent examples.[173] In the case of the latter, the United States closed hundreds of facilities of all sizes but had wished to maintain a few airfields in Iraq indefinitely. As it turned out, the United States was unable to negotiate a Status of Forces Agreement with the Iraqi government and closed all its bases.[174]

Second, path aversion can influence near-term decisions regarding how an institution trains, equips, and plans to operate its forces. After the fall of Saigon, the U.S. Army largely turned away from COIN operations, dramatically cutting special forces units and abandoning COIN training and doctrinal development in the force at large. This reflected both an institutional desire to move beyond a decade-long preoccupation with Vietnam and a U.S. Army cultural preference for large-scale conventional combat.[175] The reset was also driven by real-world demands; after a decade of neglect, U.S. Army forces in Europe were in desperate need of new operational level

---

[172] Richard Nixon, "Address to the Nation on the War in Vietnam," speech, November 3, 1969b. For Nixon's explanation of the broader national security strategy (the Nixon Doctrine) behind Vietnamization, see Richard Nixon, "Informal Remarks in Guam with Newsmen," July 25, 1969a.

[173] During the drawdown, USAF left behind ten MOBs in Vietnam. It also returned Takhli Royal Thai Air Base to Thailand on September 12, 1974, and switched Ubon Royal Thai Air Base to standby status on November 1, 1974. The two bases were no longer needed because the units that had been based there had moved to other Pacific Air Forces bases or returned to the United States as part of the broader drawdown of forward deployments associated with the U.S. withdrawal. For a list of USAF bases in Vietnam and Thailand as of December 1967, see Herman S. Wolk, *USAF Plans and Policies: Logistics and Base Construction in Southeast Asia, 1967*, Washington, D.C.: USAF Historical Division Liaison Office, 1968, p. 27. For discussion of base drawdown in Thailand, see E. R. Hartsook, *The Air Force in Southeast Asia: The End of the U.S. Involvement, 1973–1975*, Washington, D.C.: Office of Air Force History, 1980, p.68.

[174] Ewen MacAskill, "Iraq Rejects U.S. Request to Maintain Bases After Troop Withdrawal," *The Guardian*, October 21, 2011. For a comprehensive treatment of the U.S. military withdrawal from Iraq, see Rick Brennan, Jr., Charles P. Ries, Larry Hanauer, Ben Connable, Terrence K. Kelly, Michael J. McNerney, Stephanie Young, Jason H. Campbell, and K. Scott McMahon, *Ending the U.S. War in Iraq: The Final Transition, Operational Maneuver, and Disestablishment of United States Forces–Iraq*, Santa Monica, Calif.: RAND Corporation, RR-232-USFI, 2013.

[175] See Andrew F. Krepinevich, Jr., *The Army and Vietnam*, Baltimore: Johns Hopkins University Press, 1988. A broader treatment of Army culture can be found in Carl Builder, *The Masks of War: American Military Styles in Strategy and Analysis*, Baltimore: Johns Hopkins University Press, 1989.

concepts, as well as modern armored vehicles, fire support, and other conventional warfighting improvements.

Third, path aversion can trigger institutional reactions designed to limit the ability of future decision makers to go down a path that has clearly failed. For example, as will be discussed in the following section, it is commonly believed that the Army Chief of Staff, General Creighton Abrams, took steps to increase the dependence of the active force on the Army National Guard and Reserve so that a major conflict like Vietnam could not be conducted without first calling up the Guard and Reserve, steps that presumably would require substantial congressional and public support prior to initiation of major combat operations. There is no documentary evidence to support this claim about Abrams' motives, but, if true, it would exemplify path aversion at the highest levels. A more straightforward example of this type of path aversion is the 1973 War Powers Resolution limiting presidential power to use military force without congressional support. This reflected a strong congressional consensus that the executive branch had had too free a hand in decisions that led to the unhappy outcome in Vietnam.

Finally, the most visible and discussed manifestation of path aversion is the impact of a policy failure on future decisions by national leaders regarding the use of military force. This begins with the lessons that national leaders, policy elites, and the public take away from the past failure. These lessons are usually contested and can be found in competing narratives about the true causes of the failure. To some degree, these debates are never fully resolved, but they can directly affect policy in the form of new strategies or doctrines. The Nixon (Guam) Doctrine of 1969 presented a new strategy that was designed to limit the future use of U.S. ground forces when Asian security partners faced internal threats.[176] The Weinberger Doctrine of 1984 and the Powell Doctrine of 1990 (mainly a restatement of Weinberger applied to the prospective use of force against Iraq) were also reactions to the Vietnam experience, explicitly identifying six tests that must be met before the United States would commit combat forces abroad.[177] To the extent that past failures lead current leaders to use force more cautiously (at least under conditions reminiscent of the failed operation), these failures likely lead to fewer bases and facilities abroad than would have been the case absent path aversion. By way of example, if the United States had sent combat forces to fight in El Salvador's civil war in the 1980s, they likely would have quickly overwhelmed existing Salvadoran military infrastructure, requiring the expansion of current facilities and creation of new ones (particularly airfields). If that had occurred, experience elsewhere suggests a fairly high probability that USAF would still have an airfield in El Salvador.

---

[176] See Nixon, 1969a.

[177] Caspar W. Weinberger, "The Uses of Military Power," remarks at the National Press Club, Washington, D.C., November 28, 1984. See also Walter LaFeber, "The Rise and Fall of Colin Powell and the Powell Doctrine," *Political Science Quarterly*, Vol. 124, No. 1, Spring 2009, pp. 71–93, and Eric R. Alterman, "Thinking Twice: The Weinberger Doctrine and the Lessons of Vietnam," *The Fletcher Forum*, Winter 1986, pp. 93–109.

Most recently, many view the American experiences conducting COIN operations in Iraq and Afghanistan as failures, and this has led to path averting behavior in the U.S. government and public. Evidence of path aversion in the public can be found in opinion surveys, such as a 2012 New York Times/CBS News Poll that asked, "Do you think the United States is doing the right thing fighting the war in Afghanistan?" Sixty-nine percent of respondents answered "No, the U.S. should not be involved."[178]

In his 2008 campaign for president, then-Senator Barack Obama clearly embraced path aversion regarding the war in Iraq, which he described in terms reminiscent of Nixon's "Silent Majority" speech:

> We are less safe and less able to shape events abroad. We are divided at home, and our alliances around the world have been strained. The threats of a new century have roiled the waters of peace and stability, and yet America remains anchored in Iraq.[179]

Obama's remarks mirrored popular attitudes in 2008; the number of Americans who viewed the Iraq war as a mistake peaked that year at 63 percent.[180] Ending the war in Iraq became a central theme in Obama's presidential campaign:

> So when I am Commander-in-Chief, I will set a new goal on Day One: I will end this war. Not because politics compels it. Not because our troops cannot bear the burden—as heavy as it is. But because it is the right thing to do for our national security, and it will ultimately make us safer.[181]

President Obama delivered on that promise, withdrawing all U.S. forces from Iraq by December 2011 and, in the process, shedding upwards of 500 U.S. military facilities in that country.[182]

U.S. policy toward Libya in 2011 is evidence of path aversion by President Obama regarding ground-centric COIN, not just in Iraq, but in the broader Middle East. As the Libyan civil conflict heated up in 2011, the Iraq experience clearly influenced the President's decisions regarding the scale and type of U.S. involvement, in particular seeking to avoid any commitment of ground forces and reflecting the desire to have NATO allies play a leadership role.[183]

---

[178] Elisabeth Bumiller and Allison Kopicki, "Support in U.S. for Afghan War Drops Sharply, Poll Finds," *New York Times*, March 26, 2012.

[179] Barack Obama, "Obama's Speech on Iraq, March 2008," Council on Foreign Relations, March 19, 2008.

[180] Andrew Dugan, "On 10th Anniversary, 53% in U.S. See Iraq War as Mistake," Gallup, March 18, 2013.

[181] Obama, 2008.

[182] Reuters reported that at the height of the war, the United States had over 500 facilities in Iraq. See Joseph Logan, "Last U.S. Troops Leave Iraq, Ending War," Reuters, December 18, 2011.

[183] See Christopher Chivvis, *Toppling Qaddafi: Libya and the Limits of Liberal Intervention,* Cambridge, UK: Cambridge University Press, 2013, and Karl P. Mueller, ed., *Precision and Purpose: Airpower in the Libyan Civil War*, Santa Monica, Calif.: RAND Corporation, RR-676-AF, 2015.

Path aversion appears to have colored U.S. policy toward the Syrian civil war as well. The White House resisted calls from congressional critics to intervene in Syria until ISIS territorial gains in 2014 (ironically in Iraq) put the group on the doorstep of Baghdad. The risk of Baghdad falling to ISIS was sufficiently real that the United States had little choice but to intervene against ISIS in both Iraq and Syria. Path aversion, although not strong enough to prevent the intervention, nevertheless played a role, constraining U.S. military operations to participation in coalition airstrikes and train-and-advise missions for more than a year. Critics maintained that the air-only strategy would not work, calling for U.S. ground forces or, at minimum, joint terminal attack controllers (JTACs) to support Iraqi and Kurdish forces.[184] Both the White House and at least some senior military officers were reluctant to move in that direction, largely because of fears that this would ultimately lead to an open-ended, potentially large U.S. ground commitment in Iraq.

As of April 2016, the administration has reluctantly expanded operations against ISIS to include an "Expeditionary Targeting Force," more advisers, and direct support (including USMC artillery) for Iraqi forces.[185] Although there does appear to be a consensus among national leaders and the public that ISIS must be defeated, an "Iraq Syndrome" continues to exist to some degree in the public, military, and White House. The commitment of large U.S. ground forces to civil conflicts in the Middle East and North Africa appears unlikely in the near term, with the United States continuing to rely on some combination of U.S. airpower and special forces teamed with local forces. The impact of this Iraq Syndrome on posture, however, has been mixed. U.S. military posture in the Persian Gulf has remained fairly constant (at least as measured by the number of bases available to U.S. forces) due to ongoing operations in Afghanistan and is even creeping upward to support Operation Inherent Resolve against ISIS. On the other hand, it likely would have expanded much more if the United States had committed large ground formations to the conflicts in Libya and/or Syria or reintroduced them to Iraq.

The Vietnam Syndrome: A Case of Path Aversion in U.S. Foreign Policy?

Since the end of the war in Iraq, critics of American policy have revisited a phenomenon from the past to explain the Obama administration's hesitance to use conventional military force

---

[184] Regarding calls to send JTACs to Iraq, see Terry Atlas, "Obama Under Pressure to Send U.S. Target Spotters to Iraqi Front," *Bloomberg Business*, May 22, 2015. For an essay arguing that U.S. ground forces are required, see David Johnson, "Means Matter: Competent Ground Forces and the Fight Against ISIL," *War on the Rocks*, March 19, 2015. Senior USAF officers are more skeptical about the wisdom of sending JTACs to Iraq. See, for example, James Kitfield, "Gen. Carlisle: JTACS in Iraq Would Mean Lots of U.S. Ground Troops," *Breaking Defense*, July 23, 2015; and David Deptula, "How to Defeat ISIL: It's All About the Strategy," *Breaking Defense*, September 5, 2014.

[185] Barbara Starr, "Army's Delta Force Begins to Target ISIS in Iraq," CNN Politics, February 29, 2016; and Michael S. Schmidt, "Marine Base in Northern Iraq Is Confirmed by Pentagon," *New York Times*, March 21, 2016.

in Libya, Syria and Iraq.[186] Journalists and scholars have compared this hesitance to the Vietnam Syndrome, broadly defined as the unfavorable memory of American defeat in the Vietnam War that took root in the mid-1970s.[187]

The Vietnam Syndrome took form as two very different memories or lessons learned. For those who believed the war was just, but badly executed, Vietnam taught several lessons:

- Military power must be applied in a decisive manner with minimal civilian interference.
- Political and military objectives must be clear and achievable.
- A failure to apply military force effectively would create division at home and harm U.S. prestige and power globally.[188]

For those who saw the war as ill-conceived or immoral, it taught very different lessons:

- The United States lacked the wisdom, power, and moral character to intervene effectively and humanely in internal conflicts.
- The government would go to extremes to hide the ugly truths of failed policies and guerrilla warfare from the American public.
- U.S. power and prestige were greatly harmed by such interventions.[189]

Most Americans sat somewhere between these poles, embracing a hybrid set of lessons, including the recognition that guerrilla conflict in far away jungles was a particularly nasty business, that U.S. conventional military superiority did not necessarily lead to victory, and that advising and assisting a partner nation was a slippery slope that could lead to a long and costly conflict. These competing narratives all influenced elite and popular attitudes, as well as policies, over the next 40 years.[190]

---

[186] A number of writers have examined the reemergence of the Vietnam Syndrome. See Steven Conn, "Jeb Bush, Iraq, and the Vietnam Syndrome," *Huffington Post*, May 21, 2015; Leon Hadar, "Let the Iraq Syndrome Kick in," *National Interest*, May 17, 2013; Marvin Kalb, "It's Called the Vietnam Syndrome, and It's Back," Washington, D.C.: Brookings Institution, January 22, 2013; and Erik Slavin, "Decades Later, 'Vietnam Syndrome' Still Casts Doubts on Military Action," *Stars and Stripes*, November 12, 2014.

[187] Gail E. S. Yoshitani, *Reagan on War: A Reappraisal of the Weinberger Doctrine, 1980–1984*, College Station: Texas A&M University Press, 2012, p. 12.

[188] For examples of this school of thought, see Guenter Lewy, *America in Vietnam*, Oxford, UK: Oxford University Press, 1980; Mark Moyar, *Triumph Forsaken: The Vietnam War, 1954–1965*, Cambridge, UK: Cambridge University Press, 2009; and U. S. Grant Sharp, *Strategy for Defeat: Vietnam in Retrospect*, San Rafael, Calif.: Presidio Press, 1978. Also from this perspective but focused on mistakes in military strategy is Harry G. Summers, Jr., *On Strategy: A Critical Analysis of the Vietnam War*, New York: Presidio Press, 1995.

[189] For critical treatments of the Vietnam War, see David Halberstam, *The Best and the Brightest*, New York: Ballantine Books, 1993; H. R. McMaster, *Dereliction of Duty: Lyndon Johnson, Robert McNamara, the Joint Chiefs of Staff and the Lies That Led to Vietnam*, New York: Harper Collins, 1997; and Gabriel Kolko, *Anatomy of a War: Vietnam, the United States and the Modern Historical Experience*, New York: The New Press, 1994. Two critical books that focus on the Vietnam Syndrome are Michael T. Klare, *Beyond the 'Vietnam Syndrome': U.S. Interventionism in the 1980s*, Washington, D.C.: Institute for Policy Studies, 1982; and Geoff Simons, *Vietnam Syndrome: Impact on U.S. Foreign Policy*, New York: St. Martin's Press, 1998.

[190] The most comprehensive treatment of Vietnam's impact on American foreign policy elite attitudes (up to the early 1980s) is Ole Holsti and James Rosenau, *American Leadership in World Affairs: Vietnam and the Breakdown of Consensus*, London: Allen & Unwin, 1984. For more-recent assessments of the war's lasting impact, see David

For Congress, the memory of Vietnam resulted in legislation to restrain executive power to make war. For the U.S. armed forces, the Vietnam Syndrome shaped doctrine and force structure for the next generation. For the American public, the conflict in Southeast Asia violated its preferred way of fighting a war. Americans favored wars of short duration, in which the military implemented a strategically offensive doctrine to destroy an enemy with overwhelming force.[191] For the President, the specter of Vietnam resulted in new limitations on executive power and the burden of rebuilding faith in the American military and its ability to exert influence abroad. Defeat in Southeast Asia loomed over decisionmaking for roughly a decade-and-a-half, until 1991, when President George H.W. Bush proudly proclaimed, "by God, we've kicked the Vietnam Syndrome once and for all" shortly after the conclusion of Operation Desert Storm.[192]

Bush's proclamation after U.S. forces routed the Iraqi Army suggested a new confidence in the United States that the military could again exert influence around the world. Despite Bush's statement, however, the effect of the Vietnam Syndrome had deteriorated well before Operation Desert Storm, suggesting that the memory of Vietnam had lost much of its ability to constrain decisionmakers long before it was replaced with a new memory based on a recent victory.

Some observers of the Vietnam War (most notably Harry G. Summers, Jr., and historian Lewis Sorley) argue that the effects of the Vietnam Syndrome appeared before the last American troops departed in 1975. According to this view, the earliest signs emerged shortly after the return of General Abrams to the United States from his tenure as Commander of Military Assistance Command Vietnam. Abrams served as U.S. Army chief of staff starting in October 1972 and continued in this position until his death from lung cancer in September 1974. During his term as Chief of Staff, Abrams made it a priority to reform the emerging all-volunteer force. Summers and Sorley interpret Abrams's efforts to more fully integrate the reserve and active components as an effort to prevent a future president from putting troops in harm's way without careful consideration and the support of the American people.[193] According to Summers,

---

Anderson and John Ernst, *The War That Never Ends: New Perspectives on the Vietnam War*, Lexington: University Press of Kentucky, 2014, and Andrew Preston, "Rethinking the Vietnam War: Orthodoxy and Revisionism," *International Politics Reviews*, Vol. 1, September 2013, pp. 37–39.

[191] Adrian R. Lewis, *The American Culture of War: The History of U.S. Military Force from World War II to Operation Iraqi Freedom*, New York: Routledge Taylor and Francis Group, 2007, p. 8, 35–36.

[192] George H. W. Bush, "Remarks to the American Legislative Exchange Council," March 1, 1991.

[193] According to historian Gian Gentile, Harry G. Summers, Jr. was the first to claim that General Abrams shifted key support components to the reserve to prevent a future President from going to war without first calling up the reserves. See Gian Gentile and Conrad Crane, "Understanding the Abrams Doctrine: Myth Versus Reality," *War on the Rocks*, December 9, 2015. See also Harry G. Summers, Jr., "The Army After Vietnam," in Kenneth J. Hagan and William Roberts, eds., *Against All Enemies: Interpretations of American Military History from Colonial Times to the Present*, Westport, Conn.: Praeger, 1986, p. 363; and Lewis Sorley, *Thunderbolt: From the Battle of the Bulge to Vietnam and Beyond: General Creighton Abrams and the Army of His Times*, New York: Simon and Schuster, 1992, especially pp. 361–364. These issues are also discussed in Andrew J. Bacevich, *The New American Militarism: How Americans Are Seduced by War*, New York: Oxford University Press, 2005, pp. 38–39; and Yoshitani, 2012, pp. 32–33.

General Abrams hoped this return of the army to the structure it had known throughout much of the twentieth century would correct one of the major deficiencies of the American involvement in the Vietnam War—the commitment of the army to sustained combat without the explicit support of the American people as expressed by their representatives in Congress.[194]

Abrams achieved this when the Army expanded from 13 divisions to 16 without adding additional full-time personnel. To fill out combat units and support functions, the Army would have to call up reserve forces. The president and Congress would have to consider the economic and political repercussions of mobilizing men and women and pulling them away from their jobs and families.

The Summers/Sorley view has been widely accepted for decades, but historians Gian Gentile and Conrad Crane argue that there is no documentation to support this interpretation of Abrams's reorganization initiatives.[195] Sorley does present evidence that Abrams and other senior Army officers were deeply troubled by President Johnson's decision not to call up the reserves. According to Sorley, General Harold K. Johnson, then-Army Chief of Staff, came close to resigning over the decision.[196] Sorley makes a compelling case that Abrams did seek to create a total force that could not be deployed without the reserve component. Regarding Abrams' motivation, Sorley offers no evidence that Abrams sought to force future presidents to rely on popular and congressional support; he just cites Summers's earlier (and undocumented) claim. Rather, Abrams and other senior officers appeared to be focused exclusively (or at least primarily) on ensuring the U.S. Army's capabilities during wartime.

Congress also moved to suppress the president's ability to act unilaterally and to prevent any further loss of the legislative branch's constitutional powers in regards to the conduct of war. In November 1973, Congress overrode a veto by Nixon on the War Powers Resolution.[197] The resolution required the president to submit a report to Congress within 48 hours of deploying military personnel to combat areas; if Congress failed to pass a declaration of war or a resolution to allow the troops to remain, then the president would have to withdraw the deployed forces.[198] The War Powers Resolution narrowed the latitude of the president to commit troops to war or build up related basing. This was a reversal from nine years earlier, when Congress passed the

---

[194] Summers, 1986, p. 363.

[195] Historian Conrad Crane was the first to point out the lack of documentary evidence to support these claims; see Conrad C. Crane, *Avoiding Vietnam: The U.S. Army's Response to Defeat in Southeast Asia*, Carlisle, Pa.: Strategic Studies Institute, Army War College, 2002, p. 5; and Gentile and Crane, 2015.

[196] Sorley, 1992, p. 361.

[197] Public Law 93-148 is also referred to as the War Powers Act, the title of the Senate version of the bill. It is unfortunate that the final bill title replaced "Act" with "Resolution," as that makes the law sound like something less than what it is—a bill passed by both the Senate and the House of Representatives that has the full force of law (Public Law 93-148, 87 Stat. 558, Interpretation of Joint Resolution, November 7, 1973).

[198] Raymond Bonner, *Weakness and Deceit: U.S. Policy and El Salvador*, New York: Times Books, 1984, pp. 272–273; and Public Law 93-148.

Tonkin Gulf Resolution, authorizing President Johnson to use armed force in Vietnam and develop a massive basing network in Southeast Asia without a declaration of war.[199]

The Vietnam Syndrome Weakens Under President Reagan

In the second half of the 1970s, America called on its armed forces infrequently. Operations were minor affairs, typically consisting of evacuations of U.S. nationals from hostile areas or providing logistical support to allies. The next test for the U.S. military came in an ill-fated rescue attempt in April 1980. After the Iranian Revolution of 1979 ousted Mohammed Reza Shah Pahlavi, Iranian radicals seized the U.S. Embassy in Tehran, taking dozens of American diplomats hostage. President Carter ordered a rescue attempt, which had to be aborted due to equipment problems.[200]

The failure of the mission spawned criticisms of White House leadership and military readiness, particularly in the press.[201] According to political scientist Andrew Bacevich, Operation Eagle Claw did not replace the memory of Vietnam but highlighted continued problems with military effectiveness.

> The failure at Desert One did not erase the memory of that earlier war or instantly repeal its ostensible "lessons." It did, however, change the political atmospherics, persuading large numbers of Americans that any recurrence of such a calamity was simply unacceptable. Something needed to be done. And whatever that something was, the current incumbent of the Oval [Office] seemed like the wrong man to do it.[202]

Republican presidential candidate Ronald Reagan used the loss of a friendly Iranian regime, the hostage crisis, and the failed rescue attempt as evidence of weak leadership in the White House. His platform stressed a recommitment to American statecraft and to reclaiming lost confidence and leadership in international affairs. Additionally, campaign rhetoric included reverence for the American military, an element less prevalent in the Carter administration. The voting public largely agreed with Reagan's assertions and put him in the White House.[203]

Central America (El Salvador)

On taking office, Reagan emphasized the communist threat to U.S. interests, seeking to roll back its advance in Central America. Nevertheless, the American public and the media initially

---

[199] Public Law 88-408, 384 Stat. 78, To Promote the Maintenance of International Peace and Security in Southeast Asia, August 10, 1964.

[200] JCS, Special Operations Review Group, *Rescue Mission Report,* Washington, D.C., 1980.

[201] See George C. Wilson and Michael Getler, "Anatomy of a Failed Mission," *Washington Post,* April 27, 1980.

[202] Bacevich, 2005, p. 105.

[203] Bacevich, 2005, pp. 106–17. Reagan also used Carter's July 1979 "Crisis of Confidence" speech to build a case for weakness in executive leadership, rather than a weak American spirit. See Yoshitani, 2012, pp. 4–5.

opposed Reagan's efforts, and his attempts to assist the El Salvadoran COIN provide one example of American resistance to intervention in developing countries after the Vietnam War. After a coup d'état in 1931, a military regime took control in El Salvador and, over the years, responded to peasant uprisings with extreme violence. Decades of brutal treatment by the El Salvadoran military government fomented unrest among the population. Political and social turmoil pushed the country into civil war, partially influenced by the success of the Sandinistas in Nicaragua against the regime of Anastosio Somoza DeBayle in 1979. Five guerilla factions under the banner of the FMLN served as the military arm of the Revolutionary Democratic Front, the political alliance of numerous leftist groups.[204]

American assistance to the El Salvadoran Armed Forces (ESAF) started during the Carter administration; however, these efforts expanded under his successor. The Reagan administration viewed Central America through a Cold War lens and focused its attention on El Salvador to show U.S. resolve against communism in its own hemisphere. Nevertheless, the memory of Vietnam remained too fresh for the public, the media, Congress, and many in the military, forcing Reagan to accept a small American military presence in country with a restrictive role.[205] Reagan acquiesced to the congressionally imposed 55-man limit for advisers in El Salvador. Additionally, U.S. soldiers could not accompany the ESAF into combat areas. Government personnel were even careful with the terms they applied to the U.S. presence in El Salvador; the Pentagon referred to personnel in country as "trainers" rather than "advisers."[206] After more than $4 billion in aid and a 12-year military presence, the political and military situation had changed enough to seek a negotiated settlement between the FMLN and El Salvador's government in 1992. The geopolitical situation changed immensely over the course of the 1980s, which allowed U.S. objectives to shift in Central America. The fall of the Berlin Wall, followed by the collapse of the Soviet Union and the disappearance of communism in Eastern Europe, lessened American fears about Marxist movements in Central America and allowed the United States to deemphasize the importance of military victory.[207]

The limited intervention in El Salvador's civil war was strongly criticized by some military professionals. A 1988 report written by four U.S. Army lieutenant colonels (written while they were students at Harvard) questioned the American ability to fight small wars and aid in COIN operations. One of the major issues raised by the study was the negative impact of the American conceptions of war and peace on the military's ability to wage war. To these officers, the War

---

[204] Richard Duncan Downie, *Learning from Conflict: The U.S. Military in Vietnam, El Salvador, and the Drug War*, Westport, Conn.: Praeger Publishers, 1998, pp.130–131.

[205] Downie, 1998, pp. 131–132; Bonner, 1984, p. 274. Letters to Congressmen, Gallup polls, and protests across the United States indicated little support for military action in Central America. See Yoshitani, 2012, pp. 44–45.

[206] Andrew J. Bacevich, *American Military Policy in Small Wars: The Case of El Salvador*, Washington D.C.: Pergamon-Brassey's International Defense Publishers, 1988, pp. 10–11, 22; Downie, 1998, pp. 131–133; Bonner, 1984, p. 273.

[207] Downie, 1998, p. 143.

Powers Resolution stymied the efforts of those serving in El Salvador. Restricting advisers from accompanying combat troops into the field prevented them from evaluating ESAF personnel, and the 55-man limit for trainers in country was arbitrary, unsupported by analysis.[208]

Reagan's efforts to thwart communism in Central America raised the public and Congress's fears of another Vietnam. Analyst Todd Greentree asserted that the limitations placed on U.S. efforts stem directly from the memory of Vietnam. Moreover, Greentree imagines vastly expanded interventions into other Central American countries without the Vietnam Syndrome. "It is not frivolous to consider that, had it not been for Vietnam, the United States almost certainly would have invaded Nicaragua, probably in 1979 to prevent the Sandinistas from coming to power."[209] This may be one of the clearest claims on behalf of the path aversion hypothesis: that the Vietnam experience directly prevented a specific military operation because of its similarity to the earlier failure.

Opposition to intervention in Central America proved a challenge to Reagan and indicated that the American people were not yet ready to commit the military to combat operations, or at least not COIN operations in tropical terrain. Nevertheless, the small military presence in El Salvador and the billions spent on fighting the FMLN also showed the will of the President to carry out a strategy in the face of popular resistance. Reagan looked to shape public opinion, rather than allowing public opinion to shape policy.[210] Additionally, the limitations placed on the number of advisers serving in El Salvador and the number of restrictions placed on their activities did more to erode the Vietnam Syndrome than strengthen it. By enforcing stringent measures to prevent casualties or an escalation, Congress helped prevent the recollection of painful memories and outrage over the loss of soldiers in a controversial conflict.

Lebanon

The next deployment of American armed forces abroad did not meet as much congressional resistance as the military presence in El Salvador. During and after the Arab-Israeli War of 1948–1949, tens of thousands of Palestinian refugees had flocked to Lebanon. The Palestine Liberation Organization (PLO) established a headquarters in the city of Beirut in the early 1970s, took control of the encamped refugees, and launched attacks against Israel. Lebanon remained an on-and-off battleground between Lebanese forces, the PLO, Syria, Israel, and Christian militias into the early 1980s. International concern grew after the June 1982 Israeli invasion of Lebanon reached Beirut, and the Israel Defense Forces threatened to destroy the PLO headquarters.

---

[208] The report also makes the case that El Salvadoran forces received training and equipment ill-suited for COIN operations; see Bacevich, 1988.

[209] Todd Greentree, *Crossroads of Intervention: Insurgency and Counterinsurgency Lessons from Central America*, Annapolis, Md.: Naval Institute Press, 2008, p.11.

[210] Yoshitani, 2012, p. 25.

A multinational peacekeeping force, Multinational Force I, arrived in Beirut in late August, composed of French, Italian, and U.S. units. The lightly armed U.S. force (a Marine Amphibious Unit) was tasked with preventing further bloodshed during the ceasefire and overseeing the departure of PLO and some Syrian forces from Beirut. Multinational Force I finished its mission well ahead of its four-week deadline and withdrew from Lebanon. Less than a week after the force's departure, massacres by Falangist militias in two Palestinian camps prompted a request from Lebanon to stabilize the situation until the Lebanese government and the Lebanese Armed Forces could take control. Multinational Force II arrived by the end of September, again consisting of Italian, French, and U.S. troops, followed later by a small British force.[211] First, the Reagan administration contemplated several options for force size and composition and decided to send another Marine Amphibious Unit, only with heavier armaments.

Reagan faced little public or congressional resistance to deploying U.S. forces to Beirut. The majority of Americans held a favorable opinion of Israel and believed its security was threatened by instability in Lebanon. Second, Reagan promised Congress that he would abide by the War Powers Resolution and submit a report. Third, the Marine Corps was a self-sufficient force that would require only a small footprint. Finally, Reagan incorporated the Cold War paradigm and cited Soviet meddling.[212]

The mission was not, however, clearly defined or thought out, and the exposed Marines soon found themselves forced to take sides in intra-Lebanese battles. Because U.S. forces increasingly provided support, including naval gunfire, to Christian Lebanese forces, U.S. forces were viewed as combatants by other factions in the civil war, which made them the target of attacks.[213] Terrorist attacks against U.S. forces escalated, and on April 18, a car bomb exploded at the U.S. Embassy. The bomb killed 57, including 17 Americans. The situation for the Marines worsened over the summer, as terrorists used rockets and artillery against their positions at Beirut International Airport. Finally, on October 23, a suicide bomber drove a truck into a building used as a Marine headquarters and barracks, killing 241 Marines.[214]

The White House did not want the American people to lose their reemerging faith in the military or question the utility of the peacekeeping force and its presence in the Middle East. Addressing the nation in October, the President warned that a U.S. withdrawal from Lebanon could result in more instability and terrorism. He heaped praise on the military and painted the loss of those killed in the bombings as a sacrifice for American vital interests and an effort at

---

[211] Anthony McDermott and Kjell Skjelsbaek, "Introduction," in Anthony McDermott and Kjell Skjelsbaek, eds., *The Multinational Force in Beirut 1982–1984*, Miami: Florida International University Press, 1991, pp. vi, vii.

[212] Yoshitani, 2012, pp. 72–75, 87–89.

[213] David Wills, *The First War on Terrorism: Counter-Terrorism Policy During the Reagan Administration*, Lanham, Md.: Rowman and Littlefield Publishers, 2003.

[214] Michael P. Mahaney, *Striking a Balance: Force Protection and Military Presence, Beirut, October 1983*, Fort Leavenworth, Kan.: School of Advanced Military Studies, U.S. Army Command and General Staff College, 2001, pp. 1, 3, 5–36.

peace in the Middle East.[215] Nevertheless, a December Harris poll found that only 24 percent of those surveyed thought the U.S. intervention in Lebanon was "worth it," and the USMC peacekeeping force left Lebanon two months later.[216]

Grenada

Reagan also took the opportunity during his television address to provide details regarding Operation Urgent Fury, the recent U.S. invasion of the island country of Grenada and the first major military operation since the Vietnam War. Maurice Bishop, a Marxist with ties to Fidel Castro, overthrew the Grenadian government in a military coup in 1979. With the aid of Cuba, Bishop began construction of an airport that Reagan claimed "looked suspiciously suitable for military aircraft, including Soviet-built long-range bombers." A more radical element of the island's militia seized control of the government and executed Bishop. This event sparked concern in other Caribbean islands, which then sought assistance from the United States. Approximately 1,000 American citizens were on the island at the time of Bishop's execution.[217]

Reagan's decision to invade Grenada drew unusual criticism from the United Kingdom and France, and the UN Security Council approved a resolution citing the invasion as a breach of international law. (The United States vetoed the resolution.) Regardless, Operation Urgent Fury gained the support of the American people, largely because the operation achieved its objectives quickly and at relatively low cost. Reagan rallied support for the invasion by connecting it to the larger struggle against communism and a sense of duty to aid fellow Americans. Some members of Congress, however, had reservations about the operation. Most Republicans rallied behind the President, while a few Democrats questioned the morality and legality of the action. The few who opposed the invasion mounted an attempt to introduce legislation requiring that the president remove U.S. troops from Grenada, but their effort gained little traction.[218]

Despite communication and coordination problems between services, which led to defense reforms in the Goldwater-Nichols Act of 1986, the Pentagon, White House, and Capitol Hill considered the mission an overall success. Some questions arose from the public concerning the necessity of the operation, but most felt that the casualties taken were tolerable.[219]

Operation Urgent Fury marks the end of the period (1975–1983) in which presidential military options were most constrained by the Vietnam Syndrome. Although the Vietnam War still symbolizes for Americans the risks of unconventional conflict in the developing world, its

---

[215] Ronald Reagan, "Address to the Nation on Events in Lebanon and Grenada," October 27, 1983.

[216] Alan J. Vick, *Proclaiming Airpower: Air Force Narratives and American Public Opinion from 1917 to 2014,* Santa Monica, Calif.: RAND Corporation, RR-1044-AF, 2015b, p. 101.

[217] Reagan, October 27, 1983.

[218] Yoshitani, 2012, pp. 120–123.

[219] Ronald H. Cole, *Operation Urgent Fury: The Planning and Execution of Joint Operations in Grenada, 12 October–2 November 1983,* Washington D.C.: Joint History Office, Office of the Chairman of the Joint Chiefs of Staff, 1997, p. 6, 52–55.

power to constrain presidential use of military power has continued to erode. The Vietnam Syndrome likely kept U.S. involvement in internal conflicts at a minimum, but between 1984 and 1990, deployments of U.S. forces for deterrent and combat purposes expanded. In 1984, U.S. aircraft shot down Iranian fighters over the Persian Gulf; in 1986, the United States bombed Libyan targets in retaliation for a terrorist attack in Germany; in 1987–1988, the U.S. conducted combat operations against Iranian forces in the Persian Gulf as part of the "tanker war"; in 1988, U.S. forces were deployed to Honduras to deter Nicaraguan military action; in 1989, the United States invaded Panama; and in 1990, it conducted Operation Desert Shield, its largest force deployment since the Vietnam War, followed by Operation Desert Storm.

Although President George H. W. Bush may not have been entirely right that, in the Kuwaiti and Iraqi deserts, the United States had kicked the Vietnam Syndrome "once and for all," Operation Desert Storm clearly marked the end of that earlier period of self-doubt and the beginning of a new era in which U.S. military power would be used with much greater frequency and confidence.[220] This era would last roughly until 2008, when growing casualties and slow progress in creating stability in Iraq appeared to have either reinvigorated the Vietnam Syndrome or replaced it as the Iraq Syndrome. As noted above, at the time of this writing, the Iraq Syndrome can be viewed as either on hold, pending the results of current operations against ISIS, or as a powerful and continuing constraint on any U.S. recommitment of ground forces in the Middle East.

Conclusions

This preliminary assessment of path aversion suggests that the idea has utility as a conceptual or framing device that can help us understand how past policy failures constrain future policy choices. With respect to posture, we argue that path aversion may make overseas basing less "sticky" in four ways: (1) by forcing a drawdown and withdrawal of U.S. forces during an operation perceived to be failing, (2) by causing postconflict institutional changes that might reduce the size or number of overseas bases (e.g., because of a shift in focus from one mission or region to others), (3) by triggering institutional efforts to limit presidential autonomy regarding use of military force (e.g., the 1973 War Powers Resolution), and (4) by preventing the development of new posture when a military intervention under consideration is constrained or rejected because of fear that it will be a repeat of the earlier failure.

We found strong historical evidence that path aversion in Vietnam and Iraq was a key driver of posture reductions. Cascading institutional effects can be documented in the U.S. military's (particularly the U.S. Army's) aversion to COIN doctrine, training, and organization following the Vietnam War, but we were not able to identify any subsequent U.S. decisions to reduce

---

[220] Bush quote from George C. Herring, "America and Vietnam: The Unending War," *Foreign Affairs*, Winter 1991–1992.

overseas posture that were directly attributable to the loss of that war. For example, USAF left its remaining bases in Thailand in 1976 at the request of the Thai government, not because of path aversion following the Vietnam War. The War Power Resolution and active congressional oversight significantly constrained U.S. military options and posture growth in Central America. Finally, fear of repeating the Iraq experience clearly constrained U.S. choices in Libya in 2011 and Syria in 2014, but the threat from ISIS has stopped any further U.S. posture reductions in the Middle East and even caused a slight increase.

In Chapter Seven, we present study findings and recommendations.

7. Conclusions

Most DoD discussions of global posture are driven by near- to middle-term concerns, be they operational or fiscal. Long-term posture planning tends to be the exception and overwhelmingly driven by projected demands associated with current strategy and operational plans. As we discussed in Chapter Three, strategy is an important consideration in long-term posture planning but is only one of four mechanisms that influence global posture over a multidecade period. Rather, as we sought to demonstrate in Chapters Four and Five, contingent events are the primary mechanism driving major posture changes. Once these changes have occurred, path dependent processes typically lock them in for extended periods. Finally, as we discussed in Chapter Six, policy failures can, under some circumstances, create path aversion—a reluctance to engage in particular types of military operations or in problematic regions. Path aversion does have enduring power as an idea, but its effects as a policy constraint tend to be fleeting or marginal.

In this final chapter, we present the main findings of this research, make recommendations for USAF leaders, and offer some final thoughts to guide long-term posture planning.

Findings

Conflict Trends Are a Limited Guide for Long-Term Posture Planning

Recent trends in frequency, location, and type of conflict can offer insights for USAF planners regarding near-term demands that they may be called upon to meet. In contrast, conflict trends offer less utility for long-term posture planning.[221] There are two reasons that this is the case. First, the most reliable long-term trends, such as declining death rates in conflicts and decreasing duration of conflicts, are not directly salient to posture planning. Frequency, location, and type of conflict are the variables most relevant for posture planners, but long-term trends are more difficult to discern. Perhaps most salient for USAF planners, decisions by U.S. leaders to use military force are not directly influenced by conflict trends. The future might, on average, have fewer and less lethal conflicts that suggest a safer world, but U.S. interests could still be threatened in significant ways, resulting in potentially large demands on USAF and the other services. Alternatively, conflict trends overall might be on the rise, but the location and type of conflict might be perceived as either not directly harming U.S. interests or beyond the ability of the United States to resolve at reasonable cost. Under these conditions, demands on USAF might decrease, particularly if U.S. leaders embraced a more restrained foreign policy or strongly

---

[221] See Chapter Two for our analysis of conflict trends.

resisted interventions in those types of conflicts or locations. In short, conflict trends are most helpful in developing near-term posture options but have less utility in determining long-term posture requirements.

Unforeseen Crises or Conflicts Have Been the Primary Agents of Change in U.S. and USAF Global Posture

Strategy is often assumed to drive plans, programs, budgets, and global posture. Historical experience suggests, however, that contingent events are the primary change agent in global posture. Strategy changes less often, and what changes do occur are frequently in reaction to unexpected crises. The major shocks that caused large increases, decreases, or shifts in overseas military presence were largely unforeseen: the beginning of the Cold War, the Vietnam War, the end of the Cold War, and 9/11. These events triggered fundamental changes in USAF global posture (see Figure 5.5). In contrast, proactive changes in strategy (occurring before a precipitating crisis) are less common and typically cause only marginal changes in posture. For example, since the U.S. government announced a policy in 2010 to rebalance forces to the Asia-Pacific region, DoD has shifted additional military capability to the Western Pacific and plans to do more in future years. Yet, even if turmoil in the Middle East and Europe had not slowed the strategy-driven posture shift, the change envisioned was likely to be small in both absolute terms and when compared to posture changes driven by unexpected events such as 9/11. This is not to suggest that the Pacific rebalance is mistaken or of no consequence, only to note that the scale of change is not comparable to that caused by the events listed earlier. Moreover, it has proven difficult to execute strategy-driven realignments in periods of strategic ambiguity and absent a compelling and urgent need. That said, as discussed in Chapter Three, if the United States moved from its grand strategy of deep engagement to one of offshore balancing (with minimal forward presence), the posture ramifications would be huge. Changes in grand strategy of that magnitude, however, are rare and would likely only happen in response to some major stimulus—which again suggests that contingent events are first among equals as mechanisms driving posture.

USAF Posture Is Characterized by Long Periods of Stability, Punctuated by Periods of Rapid Change

As Chapter Five documented, once the United States establishes large bases in a country or region, they tend to remain for extended periods. Multiple U.S. military facilities in Asia, Europe, and the Middle East date back to the 1940s and 1950s. This is primarily due to the phenomenon of path dependence, which creates positive feedback loops to maintain and enhance posture at the individual base, country, regional, and even global levels. Path dependence had a powerful effect on USAF posture during the Cold War. Although marginal adjustments are made to posture on a routine basis, it has required a major international event to break down the stabilizing influence of path dependence and create the institutional momentum for large posture changes. These posture "shocks" happened at the beginning of the Cold War, beginning and end of the Vietnam War, end of the Cold War, and after 9/11. In the post–Cold War Era, shocks have become more common, resulting in less stability. One can see the continuity and change in USAF posture by examining the proportion of bases by region.

The majority of USAF overseas bases were in Europe for most of the Cold War, except during the peak years of the Vietnam War, when just over 50 percent of USAF bases were in Asia. With the end of the Vietnam War and the drawdown in Asia bases, Europe once again possessed the majority of USAF bases, unsurprising since the major military threat to U.S. interests was located in Central Europe. Yet, most USAF bases remained in Europe well after the Cold War ended, suggesting path dependence played a role as well. This situation held until 9/11, when USAF presence in the Middle East greatly expanded and surpassed that in Europe. Yet by 2011, Europe once again hosted the plurality of USAF bases, suggesting that USAF presence in Europe is the stable equilibrium that the system returns to after disruptive events.

In the future, we should expect more changes to USAF's posture—akin to the greater volatility of the post–Cold War era. This is likely to be the product of a number of factors, including more-varied and geographically dispersed threats, DoD efforts to rely on "lighter" facilities, and changes in collective beliefs about the legitimacy or appropriateness of U.S. bases overseas. To date, the first factor—the changing international environment and, in particular, the location of security challenges—has accounted for the majority of posture revisions. Despite DoD's emphasis on "places not bases," this approach has largely failed in practice to prevent the United States from getting locked into a particular base. Many lily pad bases seem nearly as resistant to change as MOBs. It appears that path dependence in U.S. posture has more potential to be undermined due to a shift in collective beliefs stemming from a backlash against current U.S. strategy or a more general pushback against the appropriateness of foreign bases. While the factors discussed earlier will probably lead USAF to make more changes to its posture, especially in response to future shocks, path dependence also appears likely to provide a certain amount of stability to its core base network (i.e., large bases with a continuous U.S. presence).

Path Aversion Can Constrain Posture, but Effects Tend to Be Fleeting

Path aversion, the process in which U.S. leaders, policy elites, or the public resist particular types of military operations, is a less common but nevertheless important constraint on USAF posture. It is most visible when U.S. leaders decide to draw down deployed forces and associated posture prior to the full achievement of U.S. objectives, as they did in Vietnam and Thailand (1969–1975), Somalia (1993–1994), and Iraq (2007–2011). Path aversion following the Vietnam War triggered fundamental institutional reactions as well, including the Congressional War Powers Resolution of 1973 and U.S. Army doctrinal and force structure changes.[222] Finally, path aversion following Vietnam (particularly by the Congress and public) constrained Reagan's military options during El Salvador's civil war in the 1980s, likely preventing the development

---

[222] 50 U.S.C. 1541–1548.

of posture in El Salvador that would have been necessary had U.S. combat forces directly intervened. That said, Vietnam-related path aversion seems to have softened within a few years as a constraint on Presidential use of military force outside El Salvador. For example, Carter ordered the Iran hostage rescue in 1980, and Reagan ordered ground forces into Lebanon and Grenada in 1983.

More recently, path aversion toward ground-centric COIN operations in the Middle East dominated between 2008 and 2014. Today, policymakers and the public appear torn between the desire to see an end to U.S. combat operations in the Middle East and the belief that ISIS represents a form of violent extremism that cannot be ignored. Path aversion, however, has eroded significantly since 2014, and the U.S. ground force role against ISIS has now expanded to include artillery support and special forces raids. It remains to be seen whether path aversion in the Middle East will be a powerful or lasting constraint on USAF posture. Because ground-centric COIN is most strongly associated with Iraq, path aversion along those lines has arguably increased demands on USAF posture.

Geography Remains Relevant to Long-Term Posture Planning

Our assessment of 21 distinct planning vignettes found that airfields in Europe and the Middle East were used in more operational vignettes than any other region (see Chapter Three). Airfields in North Africa were well located but lacked the infrastructure to support the missions in the vignettes. Airfields in Asia were used less frequently because relatively few Asia scenarios appeared in the vignette draw and because most airfields in Asia are too remote or widely separated to be of use across regions. This finding is consistent with the findings of a 2013 RAND study that assessed the global utility of 30 airfields across several dozen scenarios in 12 distinct regions.[223] Bases in the Mediterranean littoral (Europe, North Africa, and the Middle East) are uniquely versatile because airfields in any one of these locations can be used to meet many operational demands in all three. Bases in Europe are also critical en route stops for airlift to the Middle East and Africa.

Recommendations

This research leads to three recommendations for USAF leaders and posture planners.

Test 30-year posture plan robustness against the failure of key assumptions and across a large number of possible demands. Although the future will be a combination of the familiar and the alien, we cannot definitely determine what factors will change or to what degree. Typically, planning processes are assumption-based, making judgments about the relative probability and importance of demands and other factors. Some, perhaps many, of these assumptions are bound to be wrong. To account for the irreducible uncertainties about the future,

---

[223] Pettyjohn and Vick, 2013, pp. 30–35.

the planning process should seek to reduce the importance of assumptions by designing a posture that is robust across many alternative futures, including diverse assumptions and a wide range of demands.

Use contingent event analysis to force consideration of future demands that current planners consider implausible, unimportant, or both. Planners in 2016 cannot know with certainty what problems USAF leaders will face in 2044. They also cannot know which of these problems will be policy priorities. Two basic approaches exist that can help reduce the impact of today's cognitive and institutional biases on long-term planning: massive scenario generation and contingent event analysis. The former uses simulations to consider the implications of hundreds or thousands of possible futures. Massive scenario generation is a relatively new technique that requires significant computational power but has great potential. Contingent event analysis is a simpler and less-resource intensive technique that has significant potential for growth. We recommend that USAF long-range posture planning incorporate these or similar techniques to test future posture options across a wide range of demands.

Seek a balance between "stickiness" and agility in USAF posture. When compared to the demands of a given moment, USAF global posture has always had shortfalls and excesses. Yet when viewed over a longer time horizon, this posture has proven to be highly robust. It would be considerably less robust if it were easy to close MOBs and move permanently stationed forces based on short-term understandings of optimality and military demands. MOBs and permanently stationed forces make contributions beyond those associated with host-nation security. They also are critical to USAF global mobility en route structure. MOBs, such as Ramstein, Kadena, and Yokota, are key parts of this network. MOBs also act as hubs from which permanently deployed forces can deploy to expeditionary locations elsewhere in theater or in nearby theaters. Thus, MOBs are often key enablers for USAF deployments to a wide range of less developed facilities. The positive side of path dependence can be seen in USAF in Europe. USAF (and all military) force posture in Europe faced considerable pressure to reduce only a few years ago; now it is being expanded to meet an increasingly aggressive Russia. Path dependent processes, although by no means uniformly good, have helped enhance stability in USAF posture. That said, USAF also needs agility in posture, particularly with respect to demands in areas where enduring posture is neither politically feasible nor necessary. The ideal posture would combine sufficient stickiness to maintain enduring access in key locations and sufficient agility to both surge from the locations as needed to meet out-of-area demands and shrink back as operational demands end. The recent practice of developing small, scalable facilities with partner nations for use in exercises, training, and operational rotations is intended to increase agility but, as noted in the findings, can fall prey to stickiness as well. Creating and maintaining a posture that has elements of both stability and agility will, therefore, require active management by senior leaders.

Final Thoughts

Long-range planners and strategists recognize that future demands may be quite different from those of today, sometimes in ways that would have been unthinkable a few years earlier. They account for these demands through trend and scenario analysis. These are valuable exercises but are limited by contemporary intellectual frameworks and resource constraints that lead to a focus on the "most important" or "most plausible" challenges, as well as a host of human cognitive limitations (e.g., a preference for an orderly, understandable world; a tendency to overextrapolate current challenges into the future) that bias both consensus-based and top-down planning processes.

Although aspects of the future will be starkly different, it is simultaneously true that change by no means comes easily or uniformly. Long-range planning often neglects the forces that resist change and enable continuity in policy. Such forces can be powerful. On the one hand, they can strengthen USAF posture by making it more stable in the face of political or budgetary pressures or possibly short-sighted strategy changes. USAF likely would have lost valuable bases to short-term pressures if not for these forces, particularly the phenomenon of path dependence discussed in Chapter Five. On the other hand, path dependence can inhibit efforts to fully align posture with strategy and may undermine posture agility.

Our analysis argues that planners must account for these diverse pressures as they develop posture in support of the 30-year USAF strategy. The posture model that we developed in this research offers a means to conceptually integrate and more deeply understand how these forces of continuity and change, deliberative planning and chance impact U.S. and USAF global posture. Specifically, we account for the respective contributions of strategy, contingent events, path dependence, and path aversion as driving mechanisms for change and stability in posture. If USAF wishes to have a global presence that is robust, enduring, versatile, and agile, planners must account for and integrate all four of these factors in posture concepts, plans, and programs.

Appendix. Polya Models and Simulations

Overview of Path Dependence Concepts

This appendix has two purposes. First, for readers who are interested in a more formal description of path dependence than offered in Chapter Five, we present some basic mathematical underpinning of the concept. Second, this appendix also introduces the Polya Urn model and describes our use of the model to shed light on how path dependence may have impacted USAF global posture during various historical periods.

As introduced in Chapter Five, path dependence is a process in which the outcome of the process in any period depends on the path or the set of previous outcomes. In its most general sense, path dependence means that present and future states, actions, or decisions depend on the paths of previous states, actions, or decisions.[224] Outcomes may be exogenous variables generated by an explicit function or an endogenous choice made by strategic actors—for instance, driving on the left or right side of the road. Path dependent outcomes pervade many social and economic settings. The emergence of norms, institutions, conventions, and social customs are typical examples of path dependent processes, as are equilibria in choices over technologies, electoral processes, and even laws.[225]

The sources and causes of path dependence are the subject of debate and study in economics and political science. The literature on path dependence tends to emphasize four characteristics: openness, critical junctures, closure, and constraint, which are discussed in Chapter Five.[226] Four different self-reinforcing processes can create stasis: increasing returns, functional, legitimacy, and power.[227] The analysis presented in this appendix fits within a class of modeling commonly associated with a path dependent process derived from increasing returns. *Increasing returns* means that the more a choice is made or an action taken, the greater the benefits associated with that action.[228] As we describe in the following sections, the technique we employ allows us to input historical data on USAF basing and capture a whole history of expected potential outcomes over specified periods. This permits us to compare predicted outcomes against actual data.

---

[224] This description is offered in Page, 2006, pp. 87–115.

[225] Jenna Bednar, Scott E. Page, and Jameson L. Toole, "Revised-Path Dependence," *Political Analysis*, Vol. 20, No. 2, 2012, pp. 146–156.

[226] The related causes of path dependence include self-reinforcement, positive feedback, and increasing returns. For a discussion of the formalized nuances among these concepts, see Page, 2006.

[227] Again, for a review of these processes, see Mahoney, 2000, pp. 515–526; and Thelen, 1999, pp. 392–396.

[228] Increasing returns can create path dependence, but so too can negative externalities. Negative externalities do not offer a precise taxonomy. They can be budget constraints or cognitive biases against making certain choices. In this sense, they are akin to irrationality.

General Description of a Discrete Dynamic Process

The model used for our simulations falls within the classification of discrete dynamical systems models. A dynamical system is a process describing the evolution of a system over time. The system should incorporate the variable(s) of interest (e.g., the population of bacteria in a petri dish or the ratio of rabbits to foxes within a forest). Discrete means that the system consists of isolated points over time (e.g., hours, days, years), such that the "time" variable only takes on integer values. For example, the system evolves or advances in a sequence of time steps or periods denoted as follows: $t = 0, 1, 2, 3, \ldots, n$.

For each time period, these types of models require a time evolution rule that determines how the system changes from one period to the next. The time evolution rule is also known as the feedback mechanism. Because the system is discrete, we can write the state of the model at time t as xt. It is not uncommon for the time evolution rule to be determined by a function, f, that maps the state of the system from time period t (its input) to the next time step, $t + 1$ (its output). Alternatively stated, function f maps the "current history" into the next "outcome." Thus beginning with the initial condition, or the original state of the system, $x0$ at time $t = 0$, we can apply the function to determine the state of the system in the next time period $t = 1$ as $x1 = f(x0)$. Repeating this process by applying f to $x1$ produces $x2 = f(x1)$. Continuous iteration of the function will determine all the future states, also known as the trajectory $(x1, x2, x3, x4, \ldots, xn)$ of the system for the initial condition $x0$. In this recursive manner, the state of the system is completely determined by the function f and the initial condition of the system $x0$.

Of particular interest in many iterative dynamic systems is the long-term trajectory of the system $(x1, x2, x3, x4, \ldots, xn)$. After sufficient time, does the system converge to a given value or remain bounded within a particular range? Does it diverge to infinity? Does it bounce around or exhibit any discernable pattern? Outcomes and long-run results can vary. For instance, if the model is a representation of the population of rabbits and foxes within a given ecosystem, the state of the system at any given moment may be thought of as the ratio of rabbits to foxes. Depending on the feedback rule in the system, the long-run evolution of the system could produce extinction of both species or alternating phases in which one is more plentiful than the other.

General Polya Urn Model

The ball and urn scheme, also known as the Polya process, is the most common formal representation of path dependence.[229] It is a discrete dynamical system, as described earlier. We employ this basic model for all the simulations presented here. The common motivation for using a Polya urn process is to model the emergence of order from a seemingly random state. In

---

[229] The Polya model is named after George Pólya, the Hungarian mathematician who made contributions to combinatorics and probability theory, among other fields. Arthur, 1994, treats the Polya model in the context of economic applications.

this setup, it is the set of past events or prior outcomes that determines long-run distributions over outcomes.[230] More specifically, for systems that do converge, outcomes or events transpiring early in the process generally dictate which singular long-run distribution eventuates. These are the early critical outcomes that shape the long-run process. This is reflective of how "history matters" in a dynamic setting.

In its basic setup, the ball and urn model assumes a collection of various colored balls in an urn of infinite volume. The number of balls of various colors present in the urn at the beginning of the process ($t = 0$) represents the initial conditions of the system $x0$. In every period, a ball is randomly selected from the urn. The selection of a given ball at time t may be thought of as the outcome for that period. Depending on the color of the selected ball, other balls may be added or withdrawn from the urn, according to a given feedback rule. Because the ball is selected at random, it is the composition of the urn at each draw that determines the likelihood that a ball of a given color is selected at every period. Broadly speaking, the system may be thought of as the evolving proportion of each color of ball within the urn at any given period.

Unless otherwise noted, the primary feedback rule we implement is one of increasing returns where we replace the random ball selected from the urn and add another ball of the same color. This "+1" random draw process is repeated until the proportion of balls of each color within the urn reaches a "steady state."[231] That is, the proportion approaches a long-run limit from which it does not change. This is also referred to as "lock-in." We can label this long-run limit Z. For example, if there are only two balls in the urn at the start of the process, one red (R) and one green (G), then repeating the random draw process will eventually produce a unique distribution of red and green balls within the urn.[232] The long-run proportion to which the system eventually settles down is itself a random variable uniformly distributed between 0 and 1. This means that, if the process is repeated several times, the long-run distributions reached each time will differ. Stated differently, the long-run proportion of red and green balls could be any ratio between zero and one.

Given this dynamic arrangement, a history of outcomes can be expressed as a sequence of Rs and Gs. When the sequence is sufficiently long, adding the next outcome to the sequence (an R or a G) ceases to meaningfully alter the ratio of Rs to Gs within the sequence. By this time, the process is said to be locked-in. Figure A.1 offers a visual representation of this process. It depicts change in the percentage of red and green balls on the vertical axis and the number of draws on

---

[230] The order in which these past events transpired is not relevant at any given period to the long-run distribution. Page, 2006, p. 14, makes a distinction between *path* and *phat* dependence; the latter reflects a situation in which the outcome in any period depends on the set of outcomes and opportunities that arose in a history but not upon their order.

[231] *Steady state* is loosely defined as the long-run proportion of balls of each color in the urn for which the addition of the next ball does not change the proportion by more than a nominal amount—for instance, less than 0.001. In the limit ($n \to \infty$), the change in the proportion as each new ball is added approaches 0.

[232] This trajectory is dependent on the "+1" feedback mechanism and the initial condition ($x0$) of one red and one green ball.

the horizontal axis. From the figure it is evident that Z, the long-run distribution of red to green balls, is approximately 27 percent red to 73 percent green.[233] As also noted, repetition of the simulation would result in an entirely different long-run distribution Z.

Figure A.1. Illustration of Polya Process with One Red and One Green Ball

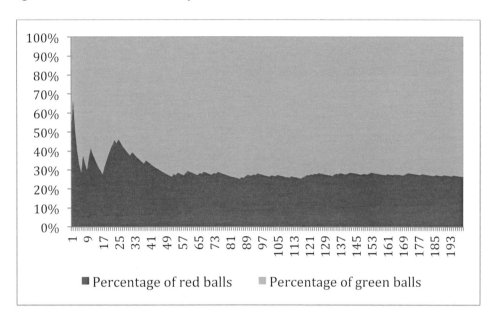

In a social science setting, Polya urn models are often employed to describe the evolution of organizational processes—for instance, how society adds new institutions or chooses technologies. The colors of the balls within the urn would then represent potential institutional outcomes or adoption of particular technological innovations. The emergence of long-run trends is thought to result as a particular outcome or set of outcomes begins to dominate the urn. This is reflective of the lock-in process defined earlier.

To add some context to the model for our setting, rather than thinking of balls of different colors in an urn, the balls may be interpreted as USAF bases in different countries or regions. These can be added to a given region, in which the various regions represent the various colors. So, for example, USAF may be allocating new facilities among three regions: Europe, Asia, and the Middle East. The number of current facilities in each of the three regions would then represent the number of balls of a particular color already in the urn. The implicit assumption of using a Polya model is that, the more infrastructure in a given area, the more likely it is to attract additional infrastructural investments, on account of increasing returns.

---

[233] From the graphic, it appears there is still small variation or blips in the long-run percentages of R and G after roughly 200 periods on the horizontal axis. If the graphic were extended to show more draws, these blips would eventually diminish to the point at which they were no longer discernable at the scale of the graphic.

Simulations

The simulations we present are based on a Polya process of multiple colors. The model we constructed allows the user to input the number of colors and the number of draws for each simulation run. The number of draws was initially set to 5,000—the number of discrete periods analyzed in every simulation. We refer to these 5,000 draws as a single run. The user may also input the initial conditions ($x0$) at the beginning of the simulation ($t = 0$). More specifically, this means the user sets the number of balls of each color within the urn before any draws have taken place. At each period, a single ball is randomly selected from the urn. A ball of the same color as the selected ball is added to the urn in each period; the randomly selected ball is also returned to the urn. The process is repeated 5,000 times, and the ratio of the colors of the different balls is captured as it changes at each turn. Eventually, the change in the ratios of colors tends toward zero as the effect from the addition of the next ball on the overall ratio of colors becomes negligible.

The 5,000 iterations of the random draw and replacement process assure that the ratio of colors within the urn converges to some value Z. This value represents the evolution of a single trajectory or the long-run distribution of colors within the urn. This trajectory is unique. This means that for any given run beginning from the same $x0$, a distinctive stable distribution of colors emerges. Since each run produces a different single distribution Z (even with the same initial conditions $x0$), we are concerned with the scope of possible trajectories—that is, the range of potential distributions for any given set of initial conditions $x0$ within which the long-run trajectories are likely to fall.[234] We therefore repeat the simulation process for each initial condition N times. This repetition generates N distinct long-run distributions: $Z1, Z2, Z3, . . . , ZN$. In the simulations presented here, $N = 10,000$.

We collect the 10,000 individual trajectories that eventuate at the conclusion of each run. This assembly of long-run distributions forms 10,000 unique data points. These data offer a broad picture of the potential range of long-term distributions that could emerge for any given initial conditions at $x0$. More specifically, they form a histogram. Because each run is independent from the next, the simulated histograms approximate normal distributions from which we can calculate means and standard deviations.[235]

The mean is calculated as

$$\overline{Z} = \frac{Z_1 + Z_2 + ... + Z_{10,000}}{10,000}$$

The standard deviation is calculated as

---

[234] To be clear, the distribution Z refers to the long-term pattern or trajectory of the eventual proportion of colors that emerges from a single run of the dynamic Polya process. This pattern is always unique for every run of the model.

[235] The precise nature of each distribution is dependent on the initial conditions, the specified feedback mechanism, and the independence of each draw.

$$\sigma = \sqrt{\frac{\sum(Z - \bar{Z})^2}{n-1}}$$

(in this case, n=10,000). The calculation of the standard deviation allows us to define expected ranges within which the majority of the long-run distributions should fall. More precisely, for a standard normal curve, roughly 95 percent of the observations fall within two standard deviations of the mean, and just over 68 percent of the observations fall within a single standard deviation from the mean. So, for any color of ball within in the urn, we are able to calculate an expected range for the long-run proportion of balls of that color for any given set of initial conditions.

Figure A.2 displays a histogram from a simulation repeated 10,000 times.

Figure A.2. Simulated Histogram

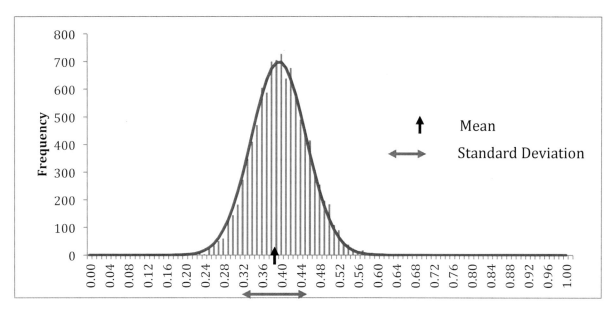

The histogram in figure A.2 has a mean of 0.395 and a standard deviation of 0.057. It is worth noting that the initial proportion of balls in the urn tends to be relatively close to the estimated means from the repeated simulations. It is the quantity of initial balls that affects the standard deviations, or the variance around these means. In general, the higher the number of initial balls of any color, the smaller the standard deviation. Therefore, the spread of possible outcomes remains highly dependent on the frequency of balls of a given color at the start of the process. It is also worth pointing out that the model will generate the same number of histograms as there are colors within the urn. So for every color, an estimated range of long-run values is produced.

Testing Path Dependence in Historical Regional USAF Basing

This section examines the degree to which path dependence may be evident in historic regional USAF data. Initially, military and policy planners might have several options among which to choose when making regional basing decisions. But after several bases or facilities of a certain type are already in place in a given area, it may be more difficult for planners to locate bases elsewhere or to transfer existing infrastructure for many reasons. Placing large facilities may create positive feedback as more facilities are placed in close proximity or increasing returns within areas may add to the benefit of locating follow-on facilities nearby.

The Polya approach is well suited for probing these types of issues because, for any given initial conditions ($x0$) and for any specified feedback rule, the model can simulate a range or dispersal of long-run values within which the observed data should be contained. If basing decisions are subject to the kinds of increasing return mechanisms highlighted above, then it is not unreasonable to expect that, over specified periods, certain basing or facilities patterns should emerge. The model helps identify the boundary conditions of these patterns. It may be the case that basing decisions are only subject to periods of "strict" path dependence for relatively short times. This type of modeling should help identify these periods and offer a firm criterion for determining the cutoff periods. If basing decisions are more flexible or even idiosyncratic, this too should be evident in the models.

Three-Region Example

In this section, we test the degree to which a path dependent process is evident in the historical share of major U.S. air bases in Europe, Asia, and the Middle East. We do so using a RAND database tracking USAF basing facilities from 1955 to 2011 in these three regions.[236] We are particularly interested in identifying the periods in which basing outcomes overseas might be appropriately described as exhibiting path dependence and the periods in which such dynamics do not hold. By using the basing data for these regions to inform the initial conditions of our Polya simulations, we can extrapolate permissible ranges within which path dependent trajectories should fall for the specified starting conditions.

Figure A.3 offers a visual depiction of the total number of major U.S. air bases across the three regions. The number of bases declined from a maximum of 74 in 1955 to a minimum of 20 in 1995. Incorporating the same data, Figure A.4 shows the distribution of these bases across the three regions as a proportion of the total number of U.S. air bases. Europe had the most air bases from 1953 to 1966, with the maximum share occurring in 1960. Asia then had the most air bases for several years, until Europe once again came out on top in 1971. Proportions remain relatively constant from the mid-1970s through the late 1980s. At this point, the end of the Cold War and

---

[236] Pettyjohn and Vick, 2013, p. 67.

beginning of the first Gulf War dramatically changed both the total number of bases and respective percentages across the three regions.

Figure A.3. Major USAF Bases Overseas, 1953–2011

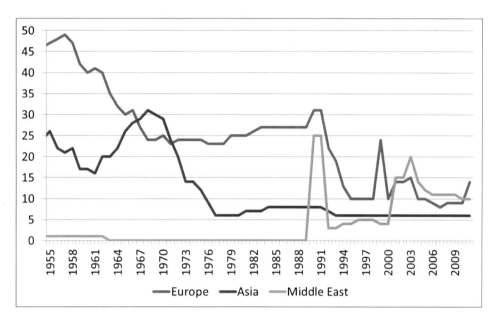

Figure A.4. Proportion of Major USAF Bases Overseas by Region, 1953–2011

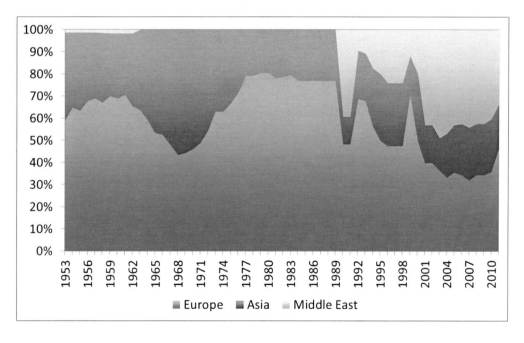

Simulation Results

We conducted two simulations that examine the degree to which USAF basing across Asia, Europe, and the Middle East conforms to the patterns that could be the result of a path dependent process. We specifically look at two distinct periods in history. The first is the post-Vietnam era, beginning in 1974 and running through the end of the Cold War in 1989. The second is the post–Cold War era up to the year 2000. The model is specifically designed to help us answer the following questions:

- Given a set of initial conditions, what is the range of long-run trajectories consistent with a basic path dependent dynamic?
- Do the basing data fit into ranges for the two specified periods?

If the data do not fit into these ranges, we can reject the premise that a general path dependent process is the primary mechanism driving regional basing outcomes. If they do fall within the outcomes allowable under path dependence, we can have more confidence that basing decisions may have a sticky component for certain periods. In this regard, the modeling exercise is a heuristic device to inform whether path dependence is an appropriate way to describe how the proportion of bases across the three regions has or has not changed.

For both of the simulations, the initial conditions are the raw number of bases in each region (Asia, Europe, and the Middle East) at the beginning of the period. As noted earlier, the simulations generate histograms depicting the range of possible long-term distributions and the frequency with which these distributions occurred in the simulation. The following analysis compares the outputs of the simulations to the actual shares of bases in the regions at the end of each period under investigation.[237]

For the first simulation, the initial conditions ($x0$) are the respective numbers of bases in three regions in 1974: 14 in Asia, 24 in Europe, and zero in the Middle East. Figure A.5 shows the simulated range for the expected share of bases in Europe. The graphic also includes markers for the mean value, the standard deviation, and an indicator (in green) reflecting the actual value of the proportion of bases in Europe at the end of the period (1989). The histogram shown in the figure follows a normal distribution with a mean of 63.2 percent and a standard deviation of 7.7 percent. In 1989, the actual share of USAF bases in 1989 in Europe was 77.1 percent. This is 13.9 percent more than the mean value produced by the Polya simulation and within two standard deviations. These results offer some limited confirmatory evidence for some type of path dependence in major air bases in Europe across this period.

Also evident in figure A.5 is the large range of outcomes consistent with path dependency. Values as low as roughly 55 percent and as high as 71 percent are within a single standard deviation of the simulated mean. Two standard deviations cover an even wider swath of proportions. This suggests that, for the initial conditions present in 1974, a broad scope of outcomes would be largely in line with a path dependent process for USAF basing in Europe.

---

[237] In subsequent simulations, we relax the constraint of comparing only the share of bases at the end of each period against the simulated means.

Figure A.5. Simulated Results for Europe

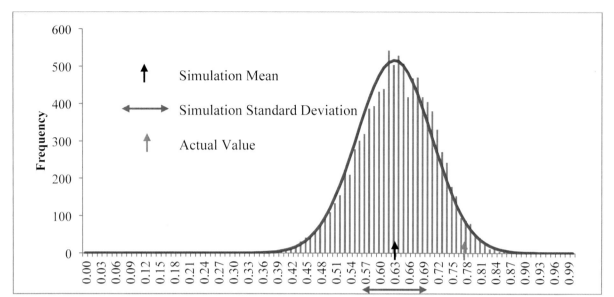

For the second simulation, the initial conditions (*x*0) are the respective number of bases in Asia, Europe, and the Middle East in 1990: four in Asia, 31 in Europe, and 25 in the Middle East. Figure A.6 shows the simulated range for the expected share of bases in the Middle East. Once again, the graphic highlights the mean value and the standard deviation and presents a marker (in green) reflecting the actual value of the proportion of bases in the Middle East at the end of the period (1989). The mean value of the histogram is 39.0 percent, with a standard deviation of the normal distribution of 6.0 percent. The actual share of bases in the Middle East in 2000 was only 20.0 percent. Not only does this value not fall within one standard deviation of the mean, it is over three standard deviations smaller than the mean. Again, Figures A.3 and A.4 make this point clear. The early 1990s show a great shakeup in the composition of U.S. air bases across all three regions.

Figure A.6. Simulated Results for the Middle East

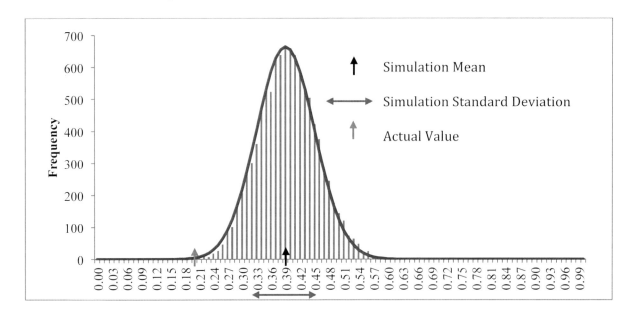

Summarizing the Simulation Results

The two preceding examples above demonstrate how using regional USAF basing data can produce expected ranges of outcomes if a path dependent progression is contributing to the data generating process. By incorporating the raw data as the initial conditions, the model offers a criterion by which we are able to judge the degree to which path dependence may be evident in USAF basing outcomes. Does the path of the data fall within the expected bandwidth of outcomes produced by the model for specified periods of time? If it does not, we are better able to reject the notion that posture decisions are largely driven by mechanisms of positive feedback and increasing returns.

This section offers a concise summarization of our model applied to each of the three regions during the following periods: 1953–1964, 1965–1973, 1974–1989, 1990–2000, 1990–2011, and 2001–2011. These periods roughly divide the post–World War II time frame by major international events affecting the United States: post–Korean War to the beginning of the Vietnam War (1953–1964), the Vietnam War era (1965–1973), the post–Vietnam Cold War period (1974–1989), the post–Cold War era up to 9/11 (1989–2000), and the period after 9/11 (2001–2011). While we have conducted the analysis across all the regions, the graphics below focus on Europe. For each period, the figures show the actual trend line in the proportion of U.S. bases (in blue) and the simulated mean (in red) from the model, as well as the one standard deviation bandwidth (in grey) around the mean. Similar to our earlier examples, the initial conditions for each simulation are the number of respective bases in the region in the first year of the simulation. The graphics offer a visual depiction for how and to what degree the actual data depart from the values the model predicts. We discuss the results for Europe, which are presented in Figures A.7 through A.11, then offer two summary tables that include the results across all three regions for each of the five historical periods.

Figure A.7 suggests that, during the period after the Korean War and prior to U.S. involvement in Vietnam, the proportion of bases in Europe remained relatively constant. Except for a brief period in 1961, the change in proportion remained within two standard deviations of the estimated mean, and for much of the period the proportion is within a single standard deviation. This steady pattern is more or less preserved throughout the Vietnam era (Figure A.8). It only began to change during the late 1970s but remained within two standard deviations of the estimated mean throughout the Cold War (Figure A.9). After the Cold War drew to a close, the proportion of European bases started to alter more drastically. Figure A.10 indicates that changes in the proportion of European bases sometimes exceeded two standard deviations from the estimated means. Ironically, by 2000, the proportion of bases in Europe returned to near where it had been in 1989. Finally, Figure A.11 shows that, since 9/11, European basing has become more fixed.

Figure A.7. Europe 1953–1964

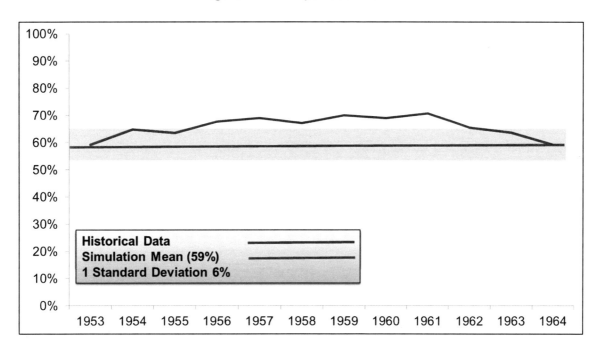

Figure A.8. Europe 1965–1973

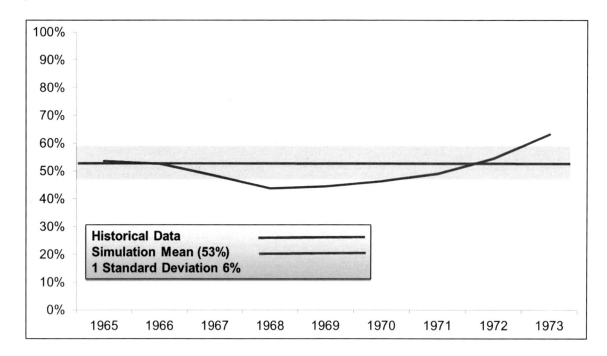

Figure A.9. Europe 1974–1989

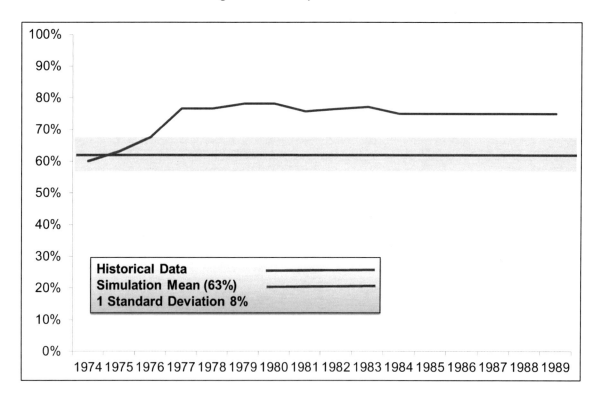

Figure A.10. Europe 1990–2000

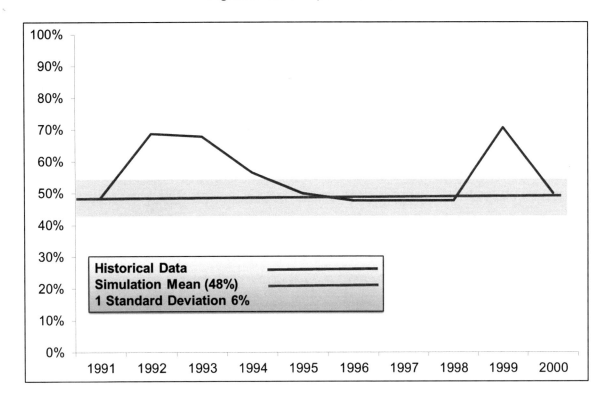

Figure A.11. Europe 2001–2011

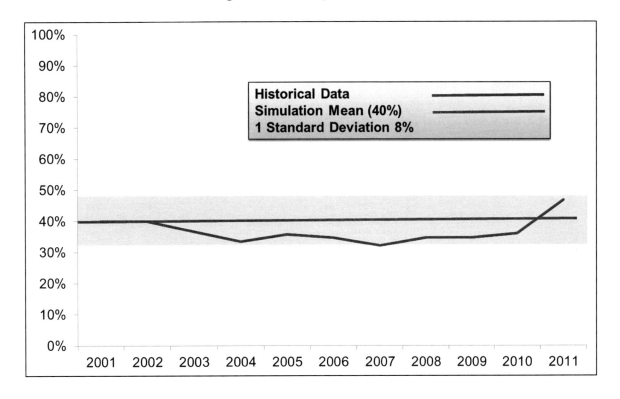

Tables A.1 and A.2 summarize results across the three regions but use different reporting standards. Table A.1 uses the endpoint, or last year of each period, as the final outcome against which estimated means are compared. Table A.2 reports the greatest departure from the simulated mean that occurred in the observed data at any year within the time frame under consideration. Regionwide, the strongest evidence of a path dependent process emerges in the data across Europe in Table A.1. But as noted above, the results for the 1990–2000 time frame are largely a consequence of similar proportions in the beginning and final years of the simulation. During this period, there were wide swings in the observed data. Across periods, both tables suggest that the post–Cold War years in fact departed the most from what could be described as path dependency. This disruptive pattern is evident across the three regions, where data consistently deviated more than two standard deviations from simulated means. In Table A.1, the post–Korean War trend falls well within the bounds that would be consistent with a path dependent process predicated on positive feedback in all three regions. But this result does not hold up in Table A.2. More recently, from 2001–2011, patterns in the proportion of regional USAF bases remained remarkably consistent in both tables. It remains to be seen if this consistency persists.

These results should be interpreted with caution. The model is designed to predict the ranges within which observed outcomes should fall under a basic path dependent process (predicated on a positive feedback mechanism). When observed outcomes fall well outside such specified ranges, we are able to confidently reject the hypothesis that path dependence is occurring. But when outcomes are within these ranges we are not able to specifically attribute these results to path dependency. We can only say that outcomes are consistent with what would generally be expected in a path dependent process. But we are unable to determine the ultimate degree to which path dependency is responsible for contributing to the observed history of outcomes.

Table A.1. Summary Evidence on Path Dependence by Region and Period (Endpoint)

	1953–1964	1965–1973	1974–1989	1990–2000	2000–2011
Europe	++	+	+	++	++
Asia	++	+	+	—	++
Middle East	+	0	0	—	+

NOTES: "++" indicates endpoint data fall within one standard deviation of simulated mean; "+" indicates endpoint data fall within two standard deviation of simulated mean; "—" indicates endpoint data fall outside of three standard deviation of simulated mean. 0 indicates insufficient observations to generate simulations.

Table A.2. Summary Evidence on Path Dependence by Region and Period (Greatest Departure)

	1953–1964	1965–1973	1974–1989	1990–2000	2000–2011
Europe	-	+	+	—	++
Asia	-	+	—	—	+
Middle East	+	0	0	—	+

NOTES: "++" indicates data fall within one SD of simulated mean; "+" indicates data fall within two SD of simulated mean; "—" indicates data fall outside of three SD of simulated mean; "-" indicates data fall outside of two SD of simulated mean. 0 indicates insufficient observations to generate simulations.

It is no great surprise that the end of the Cold War precipitated a basing shake-up for USAF overseas. But this analysis offers a sense of just how great the change in posture was. As shown here, the results for major facilities in the Middle East were well outside any expected values associated with even the widest permissible outcomes consistent with some form of path dependency.

Our experience with the Polya Urn model suggests it may have utility in future studies of USAF posture, particularly if more detailed data on posture changes were available for all facilities and for both earlier and more recent periods. Where data are available, this type of analysis may also help planners understand potential path dependencies within regions as well as across them, although, as discussed in Chapter Five, the type of feedback mechanism may differ at various levels of posture. A key challenge for future research is determining which year should be chosen for the initial conditions. This is critical because, by definition, path dependence is established early through the random choices made in the first few iterations. Thus, choices made in the late 1940s and 1950s may actually be when regional lock-in occurred, well before the earliest year (1953) for which we have data.

This relatively modest effort to model path dependence in USAF posture demonstrates that the Polya Urn approach has applicability but needs to be supported by more-detailed posture data and more-extensive qualitative work to identify a larger number of historical events that may have been triggers for path dependent processes.

Bibliography

Acemoglu, Daron, Simon Johnson, and James A. Robinson, "Reversal of Fortune: Geography and Institutions in the Making of the Modern World Income Distribution," *Quarterly Journal of Economics*, Vol. 117, No. 4, November 2002, pp. 1231–1294.

Air Force Instruction 11-2 KC-135, *Flying Operations, C/KC-135 Aircraft Configuration*, Vol. 3, March 6, 2015.

Albion, Robert G., "Distant Stations," *U.S. Naval Institute Proceedings*, Vol. 80, March 1954.

Alterman, Eric R., "Thinking Twice: The Weinberger Doctrine and the Lessons of Vietnam," *The Fletcher Forum*, Winter 1986, pp. 93–109.

Anderson, David, and John Ernst, *The War That Never Ends: New Perspectives on the Vietnam War*, Lexington: University Press of Kentucky, 2014.

Arakelian, Brian, and Rich Davison, *Redeployment of Army Forces from Germany: An Analysis of the U.S. Domestic Context*, unpublished paper, Washington, D.C.: National Defense University, 2011.

Art, Robert J., *A Grand Strategy for America*, Ithaca, N.Y.: Cornell University Press, 2003.

Arthur, Brian, *Increasing Returns and Path Dependence in the Economy*, Ann Arbor: University of Michigan Press, 1994.

Atlas, Terry, "Obama Under Pressure to Send U.S. Target Spotters to Iraqi Front," *Bloomberg Business*, May 22, 2015. As of August 14, 2015:
http://www.bloomberg.com/news/articles/2015-05-22/obama-under-pressure-to-send-u-s-target-spotters-to-iraqi-front

Bacevich, Andrew J., *American Military Policy in Small Wars: The Case of El Salvador*, Washington D.C.: Pergamon-Brassey's International Defense Publishers, 1988.

———, *The New American Militarism: How Americans Are Seduced by War*, New York: Oxford University Press, 2005.

Bapat, Navin A., "The Escalation of Terrorism: Microlevel Violence and Interstate Conflict," *International Interactions*, Vol. 40, No. 4, 2014, pp. 568–578.

Baran, Zenyo, "Fighting the War of Ideas," *Foreign Affairs*, Vol. 84, No. 6, 2005, pp. 68–78.

Bayat, Asef, "Islamism and Social Movement Theory," *Third World Quarterly*, Vol. 26, No. 6, 2005, pp. 891–908.

Beck, Colin J., "The Contribution of Social Movement Theory to Understanding Terrorism," *Sociology Compass*, Vol. 2, No. 5, 2008, pp. 1565–1581.

Bednar, Jenna, Scott E. Page, and Jameson L. Toole, "Revised-Path Dependence," *Political Analysis*, Vol. 20, No. 2, 2012, pp. 146–156.

Bennett, Andrew and Colin Elman, "Complex Causal Relations and Case Study Methods: The Example of Path Dependence," *Political Analysis*, Vol. 14, No. 3, Summer 2006.

Benson, Lawrence, *USAF Aircraft Basing in Europe, North Africa, and the Middle East, 1945–1980*, Ramstein Air Base, Germany: Headquarters, U.S. Air Forces in Europe, 1981, Declassified on July 20, 2011.

Betts, Richard K., "Is Strategy an Illusion?" *International Security*, Vol. 25, No. 2, 2000.

Blyth, Marc, *Great Transformations: Economic Ideas and Institutional Change in the Twentieth Century*, New York: Cambridge University Press, 2002.

Boas, Taylor C., "Conceptualizing Continuity and Change: The Composite-Standard Model of Path Dependence," *Journal of Theoretical Politics*, Vol. 19, No. 1, 2007.

Bonner, Raymond, *Weakness and Deceit: U.S. Policy and El Salvador*, New York: Times Books, 1984.

Brands, Hal, *The Promise and Pitfalls of Grand Strategy*, Carlisle, Pa.: U.S. Army War College, Strategic Studies Institute, August 2012.

Brennan, Jr., Rick, Charles P. Ries, Larry Hanauer, Ben Connable, Terrence K. Kelly, Michael J. McNerney, Stephanie Young, Jason H. Campbell, and K. Scott McMahon, *Ending the U.S. War in Iraq: The Final Transition, Operational Maneuver, and Disestablishment of United States Forces–Iraq*, Santa Monica, Calif.: RAND Corporation, RR-232-USFI, 2013. As of June 22, 2016:
http://www.rand.org/pubs/research_reports/RR232.html

Brooks, Stephen G., G. John Ikenberry, and William C. Wohlforth, "Don't Come Home, America: The Case Against Retrenchment," *International Security*, Vol. 37, No. 3, Winter 2012–2013, pp. 7–51.

———, "Lean Forward: In Defense of American Engagement," *Foreign Affairs*, Vol. 92, No. 130, January/February 2013.

Brunner, Ronald D., and Garry D. Brewer, *Policy and the Study of the Future: Given Complexity, Trends or Processes?* Santa Monica, Calif.: RAND Corporation, P-4912, 1972. As of June 22, 2016:
http://www.rand.org/pubs/papers/P4912.html

Builder, Carl, *The Masks of War: American Military Styles in Strategy and Analysis*, Baltimore: Johns Hopkins University Press, 1989.

Bumiller, Elisabeth, and Allison Kopicki, "Support in U.S. for Afghan War Drops Sharply, Poll Finds," *New York Times*, March 26, 2012. As of August 14, 2015:
http://www.nytimes.com/2012/03/27/world/asia/support-for-afghan-war-falls-in-us-poll-finds.html

Bush, George H. W., "Remarks to the American Legislative Exchange Council," March 1, 1991.

Calder, Kent E., *Embattled Garrisons: Competitive Base Politics and American Globalism*, Princeton, N.J.: Princeton University Press, 2007.

Camm, Frank, Lauren Caston, Alexander C. Hou, Forrest E. Morgan, and Alan J. Vick, *Managing Risk in USAF Force Planning*, Santa Monica, Calif.: RAND Corporation, MG-827-AF, 2009. As of June 22, 2016:
http://www.rand.org/pubs/monographs/MG827.html

Center for Global Development, "Mapping the Impacts of Climate Change," webpage, n.d. As of July 21, 2015:
http://www.cgdev.org/page/mapping-impacts-climate-change

Chivvis, Christopher, *Toppling Qaddafi: Libya and the Limits of Liberal Intervention*, Cambridge, UK: Cambridge University Press, 2013.

Clinton, Hillary, "America's Pacific Century," *Foreign Policy*, October 11, 2011. As of June 21, 2016:
http://foreignpolicy.com/2011/10/11/americas-pacific-century/

Cole, Ronald H., *Operation Urgent Fury: The Planning and Execution of Joint Operations in Grenada, 12 October–2 November 1983*, Washington D.C.: Joint History Office, Office of the Chairman of the Joint Chiefs of Staff, 1997.

Collier, David A., and Ruth Collier, *Shaping the Political Arena: Critical Junctures, the Labor Movement, and Regime Dynamics in Latin America*, Princeton, N.J.: Princeton University Press, 1991.

Conn, Steven, "Jeb Bush, Iraq, and the Vietnam Syndrome," *Huffington Post*, May 21, 2015. As of July 20, 2015:
http://www.huffingtonpost.com/steven-conn/jeb-bush-iraq-and-the-vie_b_7349910.html

Converse III, Elliot V., *Circling the Earth: United States Military Plans for a Postwar Overseas Military System, 1942–1948*, Maxwell Air Force Base, Ala.: Air University Press, 2005.

Cooley, Alexander, *Base Politics: Democratic Change and the U.S. Military Overseas*, Ithaca, N.Y.: Cornell University Press, 2008.

————, *Great Games, Local Rules: The New Great Power Contest in Central Asia*, Oxford, UK: Oxford University Press, 2012.

Cooley, Alexander, and Daniel Nexon, "The Empire Will Compensate You': The Structural Dynamics of the U.S. Overseas Basing Network," *Perspectives on Politics*, Vol. 11, No. 4, December 2013, pp. 1039–1042.

Cooley, Alexander, and Hendrik Spruyt, *Contracting States: Sovereign Transfers in International Relations*, Princeton, N.J.: Princeton University Press, 2009.

Crane, Conrad C., *Avoiding Vietnam: The U.S. Army's Response to Defeat in Southeast Asia*, Carlisle, Pa.: Strategic Studies Institute, U.S. Army War College, 2002.

Craven, Wesley Frank, and James Lea Cate, *The Army Air Forces in World War II:* Vol. Five, Washington, D.C.: Office of Air Force History, 1983.

Cunningham, Keith B., and Andreas Klemmer, *Restructuring the US Military Bases in Germany: Scope, Impacts, and Opportunities*, Bonn International Center for Conversion, Report 4, June 1995.

David, Paul A., "Clio and the Economics of QWERTY," *American Economic Review*, Vol. 75, No. 2, May 1985, pp. 332–337.

Davis, Paul K., Russell D. Shaver, and Justin Beck, *Portfolio-Analysis Methods for Assessing Capability Options*, Santa Monica, Calif.: RAND Corporation, MG-662-OSD, 2008. As of June 22, 2016:
http://www.rand.org/pubs/monographs/MG662.html

Davis, Paul K., Steven C. Bankes, and Michael Egner, *Enhancing Strategic Planning with Massive Scenario Generation*, Santa Monica, Calif.: RAND Corporation, TR-392, 2007. As of June 22, 2016:
http://www.rand.org/pubs/technical_reports/TR392.html

Defense Manpower Data Center, "Active Duty Military Personnel Strengths by Regional Area and by Country," data set, March 31, 2015.

Dempsey, Martin, "FY13 National Defense Authorization Budget Request from the Department of Defense," testimony before the Senate Armed Services Committee, Washington, D.C., February 15, 2012.

————, "Hearing to Receive Testimony on the Impacts of Sequestration and/or a Full–Year Continuing Resolution on the Department of Defense," testimony before the Senate Armed Services Committee, Washington, D.C., February 12, 2013.

Deptula, David, "How to Defeat ISIL: It's All About the Strategy," *Breaking Defense*, September 5, 2014. As of August 15, 2015: http://breakingdefense.com/2014/09/how-to-defeat-isil-its-all-about-the-strategy/

Dewar, James A., Carl H. Builder, William M. Hix, and Morlie H. Levin, *Assumption-Based Planning: A Planning Tool for Very Uncertain Times*, Santa Monica, Calif.: RAND Corporation, MR-114-A, 1993. As of June 22, 2016: http://www.rand.org/pubs/monograph_reports/MR114.html

DoD—*See* U.S. Department of Defense.

Downie, Richard Duncan, *Learning from Conflict: The U.S. Military in Vietnam, El Salvador, and the Drug War*, Westport, Conn.: Praeger Publishers, 1998.

Djankov, Simeon, and Marta Reynal-Querol, "The Colonial Origins of Civil War," *CESifo Area Conference: Employment and Social Protection*, Munich: CESifo, May 2007.

Dugan, Andrew, "On 10th Anniversary, 53% in U.S. See Iraq War as Mistake," Gallup, March 18, 2013. As of August 14, 2015: http://www.gallup.com/poll/161399/10th-anniversary-iraq-war-mistake.aspx

Dudziak, Mary L., *War Time: An Idea, Its History, Its Consequences*, Oxford: Oxford University Press, 2012.

"EXCLUSIVE: Sen. McCain: Conditions in the Middle East 'More Dangerous than Any Time in the Past," *CNN Press Room*, July 13, 2014. As of July 29, 2015: http://cnnpressroom.blogs.cnn.com/2014/07/13/exclusive-sen-mccain-conditions-in-the-middle-east-more-dangerous-than-any-time-in-the-past/

Federal Bureau of Investigation, *Crime in the United States: 2013*, Washington, D.C. As of August 3, 2015: https://www.fbi.gov/about-us/cjis/ucr/crime-in-the-u.s/2013/crime-in-the-u.s.-2013/offenses-known-to-law-enforcement/expanded-homicide

Fletcher, Harry R., *Air Force Bases:* Vol. 2, *Air Bases Outside of the United States of America*, Washington, D.C.: Center For Air Force History, United States Air Force, 1993.

Foreign Affairs and National Defense Division of the Congressional Research Service, *United States Foreign Policy Objectives and Overseas Military Installations*, prepared for the Committee on Foreign Relations of the United States Senate, Washington, D.C.: U.S. Government Printing Office, April 1979.

Gaddis, John Lewis, *Strategies of Containment: A Critical Appraisal of American National Security Policy During the Cold War*, New York: Oxford University Press, 2005.

Gaibulloev, Khusrav, and Todd Sandler, "Determinants of the Demise of Terrorist Organizations," *Southern Economic Journal*, Vol. 79, No. 4, 2013, pp. 774–792.

Gardner, Daniel, *Future Babble: Why Pundits Are Hedgehogs and Foxes Know Best*, New York: Plume, 2012.

Garretson, Peter, "A Range-Balanced Force: An Alternative Force Adapted to New Defense Priorities," *Air and Space Power Journal*, May–June 2013, pp. 4–29.

Gentile, Gian, and Conrad Crane, "Understanding the Abrams Doctrine: Myth Versus Reality," *War on the Rocks*, December 9, 2015. As of July 27, 2016: http://warontherocks.com/2015/12/understanding-the-abrams-doctrine-myth-versus-reality/

Gerson, Joseph, and Bruce Birchard, eds., *The Sun Never Sets: Confronting the Network of Foreign U.S. Military Bases*, Boston: South End Press, 1991.

Gholz, Eugene, Daryl G. Press, and Harvey M. Sapolsky, "Come Home, American: The Strategy of Restraint in the Face of Temptation," *International Security*, Vol. 21, No. 4, Spring 1997, pp. 5–48.

Gilens, Martin, and Benjamin I. Page, "Testing Theories of American Politics: Elites, Interest Groups, and Average Citizens," *Perspectives on Politics*, Vol. 12, No. 3, September 2014, pp. 564–581.

Gillem, Mark L., *America Town: Building the Outposts of Empire*, Minneapolis: University of Minnesota Press, 2007.

Gleditsch, Nils Petter, Steven Pinker, Bradley A. Thayer, Jack S. Levy, and William R. Thompson, "The Forum: The Decline of War," *International Studies Review*, Vol. 15, No. 3, 2013, pp. 396–419.

Goldstein, Joshua S., *Winning the War on War: The Decline of Armed Conflict Worldwide*, New York: The Penguin Group, 2011.

Goldstein, Judith, and Robert O. Keohane, *Ideas and Foreign Policy: Beliefs, Institutions and Political Change*, Ithaca, N.Y.: Cornell University Press, 1993.

Gould, Joe, "In Europe, US Army Official Favors Host-Nation Basing," *Defense News*, March 9, 2015.

Greentree, Todd, *Crossroads of Intervention: Insurgency and Counterinsurgency Lessons from Central America*, Annapolis, Md.: Naval Institute Press, 2008.

Gunzinger, Mark A., and David A. Deptula, *Toward a Balanced Combat Air Force*, Washington, D.C.: Center for Strategic and Budgetary Assessments, 2014.

Hadar, Leon, "Let the Iraq Syndrome Kick in," *National Interest*, May 17, 2013. As of July 21, 2015:
http://nationalinterest.org/commentary/let-the-iraq-Syndrome-kick-8482

Hagel, Charles, "Remarks by Secretary Hagel at the Manama Dialogue from Manama, Bahrain," December 7, 2013.

Halberstam, David, *The Best and the Brightest*, New York: Ballantine Books, 1993.

Hartsook, E. R., *The Air Force in Southeast Asia: The End of the U.S. Involvement, 1973–1975*, Washington, D.C.: Office of Air Force History, 1980.

Hatstle, Reid, *Rational Choice in an Uncertain World: The Psychology of Judgment and Decision Making*, Thousand Oaks, Calif.: Sage Publications, 2009.

Haulman, Daniel L., "Air Force Bases, 1947–1960," in Frederick J. Shaw, ed., *Locating Air Force Base Sites: History's Legacy*, Washington, D.C.: Air Force History and Museums Program, U.S. Air Force, 2004.

Henry, Ryan, "Transforming the U.S. Global Defense Posture," in Carnes Lord, ed., *Reposturing the Force: U.S. Overseas Presence in the Twenty-First Century*, Newport, R.I.: Naval War College Press, 2006.

Hernandez, Javier C., and Floyd Whaley, "Philippine Supreme Court Approves Return of U.S. Troops," *New York Times*, January 12, 2016.

Herring, George C., "America and Vietnam: The Unending War," *Foreign Affairs*, Vol. 70, No. 5, Winter 1991–1992. As of August 18, 2015:
https://www.foreignaffairs.com/articles/vietnam/1991-12-01/america-and-vietnam-unending-war

History and Research Division Directorate of Information, *Overseas Bases: A Military and Political Evaluation*, April 2, 1962.

Hoffman, Stanley, *Gulliver's Troubles, or the Setting of American Foreign Policy*, New York: McGraw-Hill, 1968.

Holling, C. S., "Understanding the Complexity of Economic, Ecological, and Social Systems," *Ecosystems*, Vol. 4, No. 5, 2001, pp. 390–405.

Holsti, Ole, and James Rosenau, *American Leadership in World Affairs: Vietnam and the Breakdown of Consensus*, London: Allen & Unwin, 1984.

Huntington, Samuel P., *The Common Defense: Strategic Programs in National Politics*, New York: Columbia University Press, 1961.

Jakes, Lara, and Dan De Luce, "Is the San Bernardino Attack the Latest in 'Crowdsourcing' Terrorism?" *Foreign Policy*, December 3, 2015. As of March 19, 2016:

http://foreignpolicy.com/2015/12/03/is-the-san-bernardino-attack-the-latest-in-crowdsourcing-terrorism/

JCS—*See* Joint Chiefs of Staff.

Jensen, Michael, "Discussion Point: The Benefits and Drawbacks of Methodological Advancements in Data Collection and Coding: Insights from the Global Terrorism Database (GTD)," National Consortium for the Study of Terrorism and Responses to Terrorism, November 25, 2013. As of June 21, 2016:
https://www.start.umd.edu/news/discussion-point-benefits-and-drawbacks-methodological-advancements-data-collection-and-coding

Jervis, Robert, *Perception and Misperception in International Politics*, Princeton, N.J.: Princeton University Press, 1976.

Johnson, David, "Means Matter: Competent Ground Forces and the Fight Against ISIL," *War on the Rocks*, March 19, 2015. As of August 14, 2015:
http://warontherocks.com/2015/03/means-matter-competent-ground-forces-and-the-fight-against-isil/

Joint Chiefs of Staff, "United States Military Requirements for Air Bases, Facilities, and Operating Rights in Foreign Territories," JCS 570/2, Reference Group 218, Combined Chief of Staff series 360 (12-9-42), November 2, 1943.

———, "Enclosure C: Overall Examination of United States Requirements for Military Bases and Base Rights," JCS 570/40, RG 218, Combined Chiefs of Staff series 360 (19-9-42), November 7, 1945.

———, Special Operations Review Group, *Rescue Mission Report*, Washington, D.C., 1980.

———, *The National Military Strategy of the United States of America: A Strategy for Today; a Vision for Tomorrow*, Washington, D.C.: Office of the Chairman of the Joint Chiefs of Staff, 2004.

Jones, Seth G., *A Persistent Threat: The Evolution of al Qa'ida and Other Salafi Jihadists*, Santa Monica, Calif.: RAND Corporation, RR-637-OSD, 2014. As of June 22, 2016:
http://www.rand.org/pubs/research_reports/RR637.html

Kadena Air Base, "Kadena Air Base History," webpage, n.d. As of June 29, 2016:
http://www.kadena.af.mil/About-Us/History/Kadena-Air-Base-History

Kahn, Herman, *The Next Two Hundred Years: A Scenario for America and the World*, New York: William Morrow and Company, 1976.

Kahneman, Daniel, *Thinking, Fast and Slow*, New York: Farrar, Straus, and Giroux, 2011.

Kalb, Marvin, "It's Called the Vietnam Syndrome, and It's Back," Washington, D.C., Brookings Institution, January 22, 2013. As of July 21, 2015:
http://www.brookings.edu/blogs/up-front/posts/2013/01/22-obama-foreign-policy-kalb

Kaplan, Lawrence S., Ronald D. Landa, and Edward J. Drea, *History of the Office of the Secretary of Defense,* Vol. V: *The McNamara Ascendancy, 1961–1965*, Washington, D.C.: Historical Office, Office of the Secretary of Defense, 2006.

Katzenstein, Peter, *The Culture of National Security, Norms, Identity, and World Politics*, New York: Columbia University Press, 1996.

Kelly, Terrence, "Stop Putin's Next Invasion Before It Starts," *U.S. News and World Report*, March 20, 2015.

Kennedy, Paul, *Grand Strategies in War and Peace*, New Haven, Conn.: Yale University Press, 1992.

Khong, Yuen Foong, *Analogies at War: Korea, Munich, Dien Bien Phu and the Vietnam Decisions of 1965*, Princeton, N.J.: Princeton University Press, 1992.

Kitfield, James, "Gen. Carlisle: JTACS in Iraq Would Mean Lots of U.S. Ground Troops," *Breaking Defense*, July 23, 2015. As of August 15, 2015:
http://breakingdefense.com/2015/07/gen-carlisle-jtacs-in-iraq-would-mean-lots-of-us-ground-troops/

Klare, Michael T., *Beyond the 'Vietnam Syndrome': U.S. Interventionism in the 1980s*, Washington, D.C.: Institute for Policy Studies, 1982.

Kolko, Gabriel, *Anatomy of a War: Vietnam, the United States and the Modern Historical Experience*, New York: The New Press, 1994.

LaFeber, Walter, "The Rise and Fall of Colin Powell and the Powell Doctrine," *Political Science Quarterly*, Vol. 124, No. 1, Spring 2009, pp. 71–93.

Layne, Christopher, "From Preponderance to Offshore Balancing: America's Future Grand Strategy," *International Security*, Vol. 22, No. 1, Summer 1997, pp. 86–124.

———, *The Peace of Illusions: American Grand Strategy from 1940 to the Present*, Ithaca, N.Y.: Cornell University Press, 2006.

Leffler, Melvyn P., "The American Conception of National Security and the Beginnings of the Cold War, 1945–48," *The American Historical Review*, Vol. 89, No. 2, April 1984.

Legro, Jeffrey W., *Rethinking the World: Great Power Strategies and International Order*, Ithaca, N.Y.: Cornell University Press, 2005.

Leheny, David, "Terrorism, Social Movements, and International Security: How Al Qaeda Affects Southeast Asia," *Japanese Journal of Political Science*, Vol. 6, No. 1, 2005, pp. 87–109.

Lemmer, George F., "Bases," in Alfred Goldberg, ed., *A History of the United States Air Force*, New York: Arno Press, 1974.

Lempert, Robert J., Steven W. Popper, and Steven C. Bankes, *Shaping the Next One Hundred Years: New Methods for Quantitative, Long-Term Policy Analysis*, Santa Monica, Calif.: RAND Corporation, MR-1626-RPC, 2003. As of June 22, 2016: http://www.rand.org/pubs/monograph_reports/MR1626.html

Lewis, Adrian R., *The American Culture of War: The History of U.S. Military Force from World War II to Operation Iraqi Freedom*, New York: Routledge Taylor and Francis Group, 2007.

Lewy, Guenter, *America in Vietnam*, Oxford, UK: Oxford University Press, 1980.

Logan, Joseph, "Last U.S. Troops Leave Iraq, Ending War," Reuters, December 18, 2011. As of August 19, 2015: http://www.reuters.com/article/2011/12/18/us-iraq-withdrawal-idUSTRE7BH03320111218

Lostumbo, Michael J., Michael J. McNerney, Eric Peltz, Derek Eaton, David R. Frelinger, Victoria A. Greenfield, John Halliday, Patrick Mills, Bruce R. Nardulli, Stacie L. Pettyjohn, Jerry M. Sollinger, and Stephen M. Worman, *Overseas Basing of U.S. Military Forces: An Assessment of Relative Costs and Strategic Benefits*, Santa Monica, Calif.: RAND, RR-201-OSD, 2013. As of June 22, 2016: http://www.rand.org/pubs/research_reports/RR201.html

Lutz, Catherine, ed., *The Bases of Empire: The Global Struggle Against U.S. Military Posts*, New York: New York University Press, 2009.

MacAskill, Ewen, "Iraq Rejects U.S. Request to Maintain Bases After Troop Withdrawal," *The Guardian*, October 21, 2011. As of August 18, 2015: http://www.theguardian.com/world/2011/oct/21/iraq-rejects-us-plea-bases

MacDonald, Paul K., and Joseph M. Parent, "Graceful Decline? The Surprising Success of Great Power Retrenchment," *International Security*, Vol. 35, No. 4, Spring 2011, p. 7–44.

Mahoney, James, "Path Dependence in Historical Sociology," *Theory and Society*, Vol. 29, No. 4, August 2000.

Mahoney, James, and Daniel Schensul, "Historical Context and Path Dependence," in Robert E. Goodwin and Charles Tilly, eds., *The Oxford Handbook of Contextual Political Analysis*, Oxford, UK: Oxford University Press, March 2006.

Mahaney, Michael P., *Striking a Balance: Force Protection and Military Presence, Beirut, October 1983*, Fort Leavenworth, Kan.: School of Advanced Military Studies, U.S. Army Command and General Staff College, 2001.

Manyin, Mark E., Stephen Daggett, Ben Dolven, Susan V. Lawrence, Michael F. Martin, Ronald O'Rourke, and Bruce Vaughn, *Pivot to the Pacific? The Obama Administration's 'Rebalancing' Toward Asia*, Washington, D.C.: Congressional Research Service, March 28, 2012.

Marion, Forrest L.," Retrenchment, Consolidation, and Stabilization, 1961–1987," in Frederick J. Shaw, ed., *Locating Air Force Base Sites: History's Legacy*, Washington, D.C.: Air Force History and Museums Program, U.S. Air Force, 2004.

Marshall, Monty G., and Benjamin R. Cole, *Global Report 2014: Conflict, Governance and State Fragility*, Vienna, Va.: Center for Systemic Peace, 2014.

Mason, David T., Mehmet Gurses, Patrick T. Brandt, and Jason Michael Quinn, "When Civil Wars Recur: Conditions for Durable Peace After Civil Wars," *International Studies Perspectives*, Vol. 12, No. 2, 2011, pp. 171–89.

McDermott, Anthony, and Kjell Skjelsbaek, "Introduction," in Anthony McDermott and Kjell Skjelsbaek, eds., *The Multinational Force in Beirut 1982–1984*, Miami: Florida International University Press, 1991.

McDougall, Walter A., *Promised Land, Crusader State: The American Encounter with the World Since 1776*, New York: Houghton Mifflin Company, 1997.

McMaster, H. R., *Dereliction of Duty: Lyndon Johnson, Robert McNamara, the Joint Chiefs of Staff and the Lies That Led to Vietnam*, New York: Harper Collins, 1997.

Mearsheimer, John J., *The Tragedy of Great Power Politics*, New York: W.W. Norton, 2001.

Michta, Andrew A., "U.S. Needs New Bases in Central Europe," *American Interest*, June 11, 2014.

Millett, Allan R., and Peter Maslowski, *For the Common Defense: A Military History of the United States of America*, New York: The Free Press, 1994.

Moyar, Mark, *Triumph Forsaken: The Vietnam War, 1954–1965*, Cambridge, UK: Cambridge University Press, 2009.

Mueller, John, *The Remnants of War*, Ithaca, N.Y.: Cornell University Press, 2004.

Mueller, John, and Mark Stewart, "Conflating Terrorism and Insurgency," *Lawfare*, February 28, 2016. As of March 24, 2016:
https://www.lawfareblog.com/conflating-terrorism-and-insurgency

Mueller, Karl P., ed., *Precision and Purpose: Airpower in the Libyan Civil War*, Santa Monica, Calif.: RAND Corporation, RR-676-AF, 2015. As of June 22, 2016:
http://www.rand.org/pubs/research_reports/RR676.html

National Consortium for the Study of Terrorism and Responses to Terrorism (START), *Global Terrorism Database*, 2014. As of July 29, 2015:
http://www.start.umd.edu/gtd

National Geospatial Intelligence Agency Defense Mapping Agency, Automated Air Facility Information File, data set. Not available to general public.

National Intelligence Council, *Global Trends 2030: Alternative Worlds*, Washington, D.C.: Office of the Director of National Intelligence, 2012. As of July 6, 2015:
http://www.dni.gov/index.php/about/organization/global-trends-2030

National Highway Traffic Safety Administration, "U.S. Department of Transportation Announces Decline in Traffic Fatalities in 2013," December 19, 2014. As of August 3, 2015:
http://www.nhtsa.gov/About+NHTSA/Press+Releases/2014/traffic-deaths-decline-in-2013.

National Security Council, "United States Objectives and Programs for National Security," Washington, D.C., April 12, 1950. As of August 9, 2016:
http://www.trumanlibrary.org/whistlestop/study_collections/coldwar/documents/pdf/10-1.pdf

Neustadt, Richard E., and Ernest R. May, *Thinking in Time: The Uses of History for Decision Makers*, New York: Free Press, 1986.

Nixon, Richard, "Informal Remarks in Guam with Newsmen," July 25, 1969a. As of June 23, 2016:
http://www.presidency.ucsb.edu/ws/?pid=2140

———, "Address to the Nation on the War in Vietnam," speech, November 3, 1969b.

Nye, Joseph, *Is the American Century Over?* Malden, Mass: Polity, 2015.

Obama, Barack, "Obama's Speech on Iraq, March 2008," Council on Foreign Relations, March 19, 2008.

Office of the Under Secretary of Defense (Comptroller), *European Reassurance Initiative Department of Defense Budget Fiscal Year (FY) 2016*, February 2015.

———, *European Reassurance Initiative Department of Defense Budget Fiscal Year (FY) 2017*, February 2016.

O'Hanlon, Michael, and Bruce Riedel, "Land Warriors: Why the United States Should Open More Bases in the Middle East," *Foreign Affairs*, July 2, 2013.

Pacific Air Forces, *Command Strategy, Projecting Airpower in the Pacific*, Hickam AFB, Hawaii, 2014.

Page, Scott E., "Path Dependence," *Quarterly Journal of Political Science*, Vol. 1, No. 1, 2006, pp. 87–115.

Pape, Robert A., *Dying to Win: The Strategic Logic of Suicide Terrorism*, New York: Random House, 2005.

———, "Empire Falls," *National Interest*, No. 99, January/February 2009.

Parker, Andrew M., Sinduja V. Srinivasan, Robert J. Lempert, and Sandra H. Berry, "Evaluating Simulation-Derived Scenarios for Effective Decision Support," *Technological Forecasting and Social Change*, Vol. 91, 2015, pp. 64–77.

Payne, Tilghman, Commander Joint Region Marianas, briefing to Guam Roundtable participants on Joint Region Marianas, September 5, 2013.

Pettyjohn, Stacie L., *U.S. Global Defense Posture, 1783–2011*, Santa Monica, Calif.: RAND Corporation, MG-1244-AF, 2012. As of June 22, 2016:
http://www.rand.org/pubs/monographs/MG1244.html

Pettyjohn, Stacie L., and Jennifer Kavanagh, *Access Granted: Political Challenges to the U.S. Overseas Military Presence, 1945–2014,* Santa Monica, Calif.: RAND Corporation, RR-1339-AF, forthcoming.

Pettyjohn, Stacie L., and Evan Braden Montgomery, "By Land and by Sea," *Foreign Affairs*, July 19, 2013. As of June 21, 2016:
https://www.foreignaffairs.com/articles/middle-east/2013-07-19/land-and-sea

Pettyjohn, Stacie L. and Alan J. Vick, *The Posture Triangle: A New Framework for U.S. Air Force Global Presence*, Santa Monica, Calif.: RAND Corporation, RR-402-AF, 2013. As of June 22, 2016:
http://www.rand.org/pubs/research_reports/RR402.html

Pew Research Center, "Global Opposition to U.S. Surveillance and Drones, but Limited Harm to America's Image," July 2014. As of June 23, 2016:
http://www.pewglobal.org/2014/07/14/global-opposition-to-u-s-surveillance-and-drones-but-limited-harm-to-americas-image/

Piazza, James A., "Incubators of Terror: Do Failed and Failing States Promote Transnational Terrorism?" *International Studies Quarterly*, Vol. 52, No. 3, 2008, pp. 469–488.

Pierson, Paul, "Not Just What, But When: Timing and Sequence in Political Processes," *Studies in American Political Development*, Vol. 14, Spring 2000a, pp. 72–92.

———, "Increasing Returns, Path Dependence, and the Study of Politics," *American Political Science Review*, Vol. 94, No. 2, June 2000b, pp. 251–267.

————, *Politics in Time: History, Institutions, and Social Analysis*, Princeton, N.J.: Princeton University Press, 2004.

Pinker, Steven, *The Better Angels of Our Nature: Why Violence Has Declined*, New York: The Penguin Group, 2011.

Porter, Patrick, *The Global Village Myth: Distance, War and the Limits of Power*, Washington, D.C.: Georgetown University Press, 2015.

Posen, Barry R., "The Case for Restraint," *American Interest*, Vol. 3, No. 1, November/December 2007, pp. 7–17.

————, *Restraint: A New Foundation for U.S. Grand Strategy*, Ithaca, N.Y.: Cornell University Press, 2014.

Posen, Barry, and Andrew L. Ross, "Competing Visions for Grand Strategy," *International Security*, Vol. 21, No. 3, Winter 1996–1997, pp. 5–53.

Preble, Christopher A., *Power Problem: How American Military Dominance Makes Us Less Safe, Less Prosperous and Less Free*, Ithaca, N.Y.: Cornell University Press, 2009.

Preble, Christopher, and John Mueller, *A Dangerous World? Threat Perception and U.S. National Security*, Washington, D.C.: CATO Institute, 2014.

Preston, Andrew, "Rethinking the Vietnam War: Orthodoxy and Revisionism," *International Politics Reviews*, Vol. 1, September 2013, pp. 37–39.

Public Law 88-408, 384 Stat. 78, To Promote the Maintenance of International Peace and Security in Southeast Asia, August 10, 1964.

Public Law 93-148, 87 Stat. 558, Interpretation of Joint Resolution, November 7, 1973.

"RAF Mildenhall: Business Concerns as the USAF to Leave," British Broadcasting Corporation, January 9, 2015.

Ratner, Ely, *Resident Power: Building a Politically Sustainable U.S. Military Presence in Southeast Asia and Australia*, Washington, D.C.: Center for New American Security, October 2013.

Reagan, Ronald, "Address to the Nation on Events in Lebanon and Grenada," October 27, 1983.

Redman, Charles, and Ann Kinzig, "Resilience of Past Landscapes: Resilience Theory, Society, and the Longue Durée," *Conservation Ecology*, Vol. 7, No. 1, 2003. As of July 10, 2015: http://www.ecologyandsociety.org/vol7/iss1/art14/

Robson, Seth, "New Bases in Bulgaria, Romania Cost the US Over $100M," *Stars and Stripes*, October 17, 2009.

Ross, Robert S., "The Problem with the Pivot: Obama's New Asia Policy Is Unnecessary and Counterproductive," *Foreign Affairs*, November/December 2012. As of July 11, 2016: https://www.foreignaffairs.com/articles/asia/2012-11-01/problem-pivot

Rumsfeld, Donald, "Prepared Testimony of the U.S. Secretary of Defense Donald H. Rumsfeld Before the Senate Armed Services Committee: Global Posture," September 23, 2004, p. 4. As of August 17, 2015: http://www.dod.mil/dodgc/olc/docs/test04-09-23Rumsfeld.pdf

Saab, Bilal Y., "Asia Pivot, Step One: Ease Gulf Worries," *National Interest*, June 20, 2013. As of June 21, 2016: http://nationalinterest.org/commentary/asia-pivot-step-one-ease-gulf-worries-8626

Saab, Bilal Y., and Barry Pavel, *Artful Balance: Future U.S. Defense Strategy and Posture in the Gulf*, Washington, D.C.: Atlantic Council, March 2015.

Sageman, Marc, *Leaderless Jihad: Terror Networks in the 21st Century*, Philadelphia: University of Pennsylvania Press, 2008.

Sandars, Christopher, *America's Overseas Garrisons: The Leasehold Empire*, New York: Oxford University Press, 2000.

Sarkees, Meredith Reid, and Frank Wayman, *Resort to War: 1816–2007,* Washington, D.C: CQ Press, 2010. As of July 6, 2016: http://www.correlatesofwar.org/data-sets/COW-war

Schmidt, Michael S., "Marine Base in Northern Iraq Is Confirmed by Pentagon," *New York Times*, March 21, 2016. As of March 30, 2016: http://www.nytimes.com/2016/03/22/us/politics/marine-base-in-northern-iraq-is-confirmed-by-pentagon.html?_r=0

Schmitt, Eric, and Steven Lee Myers, "U.S. Is Poised to Put Heavy Weaponry in Eastern Europe," *New York Times*, June 13, 2015.

Schweller, Randall L., *Maxwell's Demon and the Golden Apple: Global Discord in the New Millennium*, Baltimore: Johns Hopkins University Press, 2014.

Shapiro, Jacob, *The Terrorist's Dilemma: Managing Violent Covert Organizations*, Princeton, N.J.: Princeton University Press, 2013.

Sharp, U. S. Grant, *Strategy for Defeat: Vietnam in Retrospect*, San Rafael, Calif.: Presidio Press, 1978.

Silver, Nate, *The Signal and the Noise: Why So Many Predictions Fail—But Some Don't*, New York: Penguin Press, 2012.

Simons, Geoff, *Vietnam Syndrome: The Impact on U.S. Foreign Policy*, New York: Palgrave Macmillan, 1998.

Slavin, Erik, "Decades Later, 'Vietnam Syndrome' Still Casts Doubts on Military Action," *Stars and Stripes*, November 12, 2014. As of July 20, 2015:
http://www.stripes.com/news/special-reports/vietnam-at-50/decades-later-vietnam-Syndrome-still-casts-doubts-on-military-action-1.313846

Sorley, Lewis, *Thunderbolt: From the Battle of the Bulge to Vietnam and Beyond: General Creighton Abrams and the Army of His Times*, New York: Simon and Schuster, 1992.

Sprout, Harold Hance, and Margaret Tuttle Sprout, *The Rise of American Naval Power, 1776–1918*, Annapolis, Md.: Naval Institute Press, 1990 (originally published by Princeton University Press, 1939, 1966).

Starr, Barbara, "Army's Delta Force Begins to Target ISIS in Iraq," CNN Politics, February 29, 2016. As of March 3, 2016:
http://www.cnn.com/2016/02/29/politics/pentagon-army-target-isis-iraq/

Stewart, Cameron, "Go-Slow Signaled on Army Build Up," *The Australian*, November 14, 2012.

Stillion, John, *Trends in Air-to-Air Combat: Implications for Future Air Superiority*, Washington, D.C.: Center for Strategic and Budgetary Assessments, 2015.

Summers, Harry G., Jr., "The Army After Vietnam," in Kenneth J. Hagan and William Roberts, eds., *Against All Enemies: Interpretations of American Military History from Colonial Times to the Present*, Westport, Conn.: Praeger, 1986, pp. 361–374.

———, *On Strategy: A Critical Analysis of the Vietnam War*, New York: Presidio Press, 1995.

Sutter, Robert G., Michael E. Brown, and Timothy J. A. Adamson, *Balancing Acts: The U.S. Rebalance and Asia-Pacific Stability*, Washington, D.C.: Elliott School of International Affairs, George Washington University, August 2013.

Taleb, Nassim Nicholas, *The Black Swan: The Impact of the Highly Improbable*, New York: Random House, 2010.

Tavris, Carol, and Elliot Aronson, *Mistakes Were Made (But Not by Me): Why We Justify Foolish Beliefs, Bad Decisions and Hurtful Acts*, New York: Harcourt, 2007.

Taylor, Rob, "Australia Embraces Marine Presence in Darwin," *Wall Street Journal*, August 14, 2014.

Tetlock, Philip E., *Expert Political Judgment: How Good Is It? How Can We Know?* Princeton, N.J.: Princeton University Press, 2005.

Thelen, Kathleen, "Historical Institutionalism in Comparative Politics," *Annual Review of Political Science*, Vol. 2, 1999.

———, "How Institutions Evolve: Insights from Comparative Historical Analysis," in James Mahoney and Dietrich Rueschemeyer, eds., *Comparative Historical Analysis in the Social Sciences*, Cambridge, UK: Cambridge University Press, 2003.

UCDP—*See* Uppsala Conflict Data Program.

United Nations, "Member States," webpage, n.d. As of August 17, 2016:
http://www.un.org/en/member-states/

———, *World Population 2012,* wall chart, New York: United Nations Population Division, 2012. As of July 27, 2015:
http://www.un.org/en/development/desa/population/publications/pdf/trends/WPP2012_Wallchart.pdf

Uppsala Conflict Data Program, "Battle-Related Deaths Dataset," Vol. 5., Uppsala University, 2015. As of June 23, 2016:
http://www.ucdp.uu.se

U.S. Air Force, *America's Air Force: A Call to the Future*, Washington, D.C.: Headquarters U.S. Air Force, 2014a.

———, *Air Force Strategic Environment Assessment: 2014–2034*, Washington, D.C.: Headquarters U.S. Air Force, 2014b.

———, *Strategic Master Plan Executive Summary*, Washington, D.C.: Headquarters U.S. Air Force, 2015.

U.S. Army Europe, "U.S. Army European Activity Set," n.d. As of June 24, 2016:
http://www.eur.army.mil/jmtc/exercises/CombinedResolve/EAS_Fact_Sheet.pdf

———, "21st TSC MK Air Base Fact Sheet," November 3, 2014.

U.S. Code, Title 50, War and National Defense, Chapter 33, War Powers Resolution, Sections 1541–1548. As of August 8, 2016:
https://www.law.cornell.edu/uscode/text/50/chapter-33

U.S. Department of Defense, *Department of Defense Base Realignment Policy*, 1978.

———, *Strengthening U.S. Global Defense Posture Report to Congress*, Washington, D.C.: September 2004.

———, *Department of Defense's Fuel Contracts in Kyrgyzstan*, Washington, D.C.: Report of the Majority Staff, Subcommittee on National Security and Foreign Affairs, Committee on Oversight and Government, December 2010.

———, *Sustaining U.S. Global Leadership: Priorities for 21st Century Defense*, Washington, D.C., January 2012.

U.S. Department of the Navy, "Report to Congress on Camp Lemonnier, Djibouti Master Plan," Washington, D.C.: August, 24, 2012.

U.S. Embassy Bucharest, "U.S. Military Engagements to Romania," May 27, 2008.

U.S. European Command, *EUCOM Provides Update on the European Reassurance Initiative*, April 20, 2015.

U.S. House of Representatives, *Mystery at Manas: Strategic Blind Spots in the Department of Defense's Fuel Contracts in Kyrgyzstan*, Washington, D.C.: Report of the Majority Staff, Subcommittee on National Security and Foreign Affairs, Committee on Oversight and Government, December 2010.

U.S. Senate, *The Gulf Security Architecture: Partnership with the Gulf Cooperation Council*, majority staff report prepared for the Committee on Foreign Relations, Washington, D.C.: U.S. Government Printing Office, 2012.

USAF—*See* U.S. Air Force.

Vandiver, John, "US Army's Last Tanks Depart from Germany," *Stars and Stripes*, April 4, 2013.

Vaughn, Bruce, and Thomas Lum, "Australia: Background and U.S. Relations," Washington, D.C.: Congressional Research Service, December 14, 2015.

Vick, Alan J., *Air Base Attacks and Defensive Counters: Historical Lessons and Future Challenges*, Santa Monica, Calif.: RAND Corporation, RR-968-AF, 2015a. As of June 22, 2016:
http://www.rand.org/pubs/research_reports/RR968.html

———, *Proclaiming Airpower: Air Force Narratives and American Public Opinion from 1917 to 2014*, Santa Monica, Calif.: RAND Corporation, RR-1044-AF, 2015b. As of June 22, 2016:
http://www.rand.org/pubs/research_reports/RR1044.html

Vine, David, *Island of Shame: The Secret History of the U.S. Military Base on Diego Garcia*, Princeton, N.J.: Princeton University Press, 2012.

Walker, Brian H., Stephen R. Carpenter, Johan Rockstrom, Anne-Sophie Crépin, and Garry D. Peterson, "Drivers, 'Slow' Variables, 'Fast' Variables, Shocks, and Resilience," *Ecology and Society*, Vol. 17, No, 30, 2012. As of July 10, 2015:
http://dx.doi.org/10.5751/ES-05063-170330

Walker, Brian H., and David Salt, *Resilience Thinking: Sustaining Ecosystems and People in a Changing World*, Washington, D.C.: Island Press, 2006.

Wallensteen, Peter, *Understanding Conflict Resolution: War, Peace and the Global System*, London: Sage, 2011.

Walt, Stephen M., *Taming American Power: The Global Response to U.S. Primacy*, New York: W.W. Norton, 2006.

Walter, Barbara F., "Why Bad Governance Leads to Repeat Civil War." *Journal of Conflict Resolution*, Vol. 59, No. 7, October 2015.

Washington, George, "Washington's Farewell Address," 1796.

Weinberger, Caspar W., "The Uses of Military Power," remarks prepared for delivery to the National Press Club, Washington, D.C., November 28, 1984. As of August 19, 2015: http://www.pbs.org/wgbh/pages/frontline/shows/military/force/weinberger.html

Wendt, Alexander, *Social Theory of International Politics*, Cambridge, UK: Cambridge University Press, 1999.

Widlok, Thomas, Anne Aufgebauer, Marcel Bradtmöller, Richard Dikau, Thomas Hoffmann, Inga Kretschmer, Konstantinos Panagiotopoulos, Andreas Pastoors, Robin Peters, Frank Schäbitz, Manuela Schlummer, Martin Solich, Bernd Wagner, Gerd-Christian Weniger, and Andreas Zimmermann, "Towards a Theoretical Framework for Analyzing Integrated Socio-Environmental Systems," *Quaternary International*, Vol. 274, 2012, pp. 259–272.

Wills, David, *The First War on Terrorism: Counter-Terrorism Policy During the Reagan Administration*, Lanham, Md: Rowman and Littlefield Publishers, 2003.

Wilson, George C., and Michael Getler, "Anatomy of a Failed Mission," *Washington Post*, April 27, 1980.

Wolk, Herman S., *USAF Plans and Policies: Logistics and Base Construction in Southeast Asia, 1967*, Washington, D.C.: USAF Historical Division Liaison Office, 1968.

Worden, Mike, *Rise of the Fighter Generals: The Problem of Air Force Leadership, 1945–1982*, Maxwell AFB, Ala.: Air University Press, March 1998.

World Bank, "GDP per Capita (current US$)," webpage, n.d. As of July 6, 2016: http://data.worldbank.org/indicator/NY.GDP.PCAP.CD

Yeo, Andrew, "Local-National Dynamics and Framing in South Korean Anti-Base Movements," *Kasarinlan: Philippine Journal of Third World Studies*, Vol. 21, No. 2, 2006.

———, "Not in Anyone's Back Yard: The Emergence and Identity of Transnational Anti-Base Network," *International Studies Quarterly*, Vol. 53, No. 3, September 2009, pp. 571–594.

———, *Activists, Alliances, and Anti-U.S. Base Protests*, Cambridge, UK: Cambridge University Press, 2011.

Yoshitani, Gail E. S., *Reagan on War: A Reappraisal of the Weinberger Doctrine, 1980–1984*, College Station: Texas A&M University Press, 2012.

Young, Joseph K., and Laura Dugan, "Survival of the Fittest: Why Terrorist Groups Endure," *Perspectives on Terrorism*, 2014, Vol. 8, No. 2, pp. 1–23.

Zakaria, Fareed, *The Post-American World: Release 2.0*, New York: W.W. Norton & Company, 2011.

Zaller, John R., *The Nature and Origins of Mass Opinion*, Cambridge, UK: Cambridge University Press, 1992.